THE EINSTEIN FACTOR

THE EINSTEIN FACTOR

A Proven New Method for Increasing Your Intelligence

**WIN WENGER, PH.D.
AND RICHARD POE**

GRAMERCY BOOKS • NEW YORK

This 2004 edition is published by Gramercy Books, an imprint of Random House Value Publishing, a division of Random House, Inc., New York, by arrangement with Crown Publishers, a division of Random House, Inc.

Random House
New York • Toronto • London • Sydney • Auckland
www.randomhouse.com

All figures by Russ Adams, including adaptations from other sources.

Printed and bound in the United States

Library of Congress Cataloging-in-Publication Data

Wenger, Win.
The Einstein factor : a proven new method for increasing your intelligence / Win Wenger and Richard Poe
p. cm.
Originally published: Rocklin, CA : Prima Pub., c1996.
Includes bibliographical references and index.
ISBN 0-517-22320-1
1. Intellect. 2. Success. I. Poe, Richard, 1958- II. Title.
BF431.W455 2004
153.9—dc22 2004040521

10 9 8 7 6 5 4 3 2 1

To the memory of Leo Zajac

CONTENTS

PREFACE

Can you increase your intelligence? Absolutely. Most people are surprised to learn that intelligence quotient (IQ) scores can be raised. Yet, few experts would deny that this is possible. The accumulated evidence has simply become too strong. To give just a few examples: Children enrolled in the Head Start program—launched in 1964 to help underprivileged preschoolers—experienced IQ gains as high as 10 points after only a few months in the program; children with learning disabilities subjected to electroencephalogram (EEG) biofeedback therapy have reportedly gained 10 to 23 IQ points, following treatment;[1] scientists at the University of California at Irvine raised the IQ scores of test subjects by 8 to 9 points, simply by having them listen to 10 minutes of Mozart's *Sonata for Two Pianos in D Major,* K.448.

Of course, such sudden and dramatic gains tend to be short-lived. IQ gains made by Head Start preschoolers, for example, tended to dissipate between the third and sixth grade.[2] The jump experienced by the Mozart listeners in Irvine faded after only 15 minutes. Educators call this the Fade-Out Effect.

Some experts argue that the Fade-Out Effect proves that you're pretty much stuck with the IQ you had at birth.

This is a little bit like saying that Wilbur and Orville Wright should have given up after their maiden flight because they managed to stay airborne only 12 seconds. What was impressive about the Wright brothers' flight was not its brevity, but the fact that it happened at all. Once the possibility of flight was established, extending its duration was a mere technicality. Indeed, it was only four years later that the Wright brothers delivered to the U.S. Army Signal Corps an airplane capable of flying 125 miles.

Today, accelerative learning researchers are in a position much like that of the Wright brothers at Kitty Hawk in 1903. Everyone knows that human intelligence can be enhanced. The challenge now is to enhance it more intensely and for longer periods of time. The techniques in *The Einstein Factor* will help you do just that.

Here, you will find some of the most effective and up-to-date methods available for releasing your Einstein Factor—that secret trigger for ingenious thought that lies within all of us. Some of the highlights included in this book are innovations such as Image Streaming, PhotoReading, Freenoting, and Borrowed Genius.

IQ is only one measure of intelligence and not necessarily the best. While experiments have shown that some of these techniques will indeed raise your IQ, as a whole they will contribute toward a more important goal: elevating your overall brainpower and enhancing memory, creativity, reading speed, specific talents, and general brain health.

The Einstein Factor is the culmination of 25 years of research in accelerative learning. Through the Project Renaissance seminars, thousands have already been helped by these techniques. We hope that in these pages you, too, will find a pathway to insight and achievement that you never dreamed possible.

Win Wenger, Ph.D.
and Richard Poe

ACKNOWLEDGMENTS

The Project Renaissance program for accelerated learning—which forms the subject matter of this book—has arisen over the last 25 years from the theories, experiments and efforts of literally thousands of brain/mind pioneers, including the hundreds of good people who participated in my thinktanks, bringing with them a multitude of special qualities and unique personal insights.

Among these intrepid thinktankers, I must cite with special gratitude Vercille Bennett, Burton Linne, Hal and Marilyn Shook of Life Management Services, Abe Goldblatt, William Schulz, Bill Webster, Dan Knight of High Performance Associates in Alexandria, Virginia, Charlotte Ward of Accelerated Learning of Maryland, John Robb, Ted Woynicz, May Leiner, Tom Gregory, Kate Jones, Joel Hamaker and the late Leo Zajac, to whose living memory I herewith dedicate this book. For every one I mention here there are ten others who should be mentioned. I ask the indulgence of those whose valuable contributions I have not been able to acknowledge by name.

Beyond the thinktankers, a number of other people have made valuable contributions to this writer's life and to this present work. I give special thanks to the following:

- Dr. Sidney J. Parnes and Beatrice Parnes, for their pioneering role in helping to develop the original Osborn-Parnes creative solution-finding system and the new Visionizing system, from which so many of us have drawn inspiration, and for their warm support, advice, and wisdom over the years, which have made such a huge difference for this writer and his work.
- Lynne Schroeder and Sheila Ostrander, the authors of *Superlearning* and *Superlearning 2000,* who opened the door for accelerated learning when so many of us didn't even know there was a door to be opened. It is they who have held the light and led the way for so many years, for all of us who advance into the unknown.
- Susan Wenger, whose brilliant wit and humor, extraordinary patience, editing suggestions, and above all ingenious and far-ranging ideas have contributed more than I can say to the creation of *The Einstein Factor* and the program it contains.
- Michael and Mary Colosi, whose material support helped keep this boat in the water.
- Paul Scheele and his team, who responded so positively to this writer's initial, tentative musings, making possible a virtual explosion of synergy, teamwork, new methods, and unprecedented breakthroughs in accelerated learning.
- Stu Cart, warm supporter, contributor, and fellow protagonist at many levels.
- Physics professor Charles P. Reinert, whose lonely and courageous measurement studies on the effects of Image Streaming were the first to help "legitimize" this program in the eyes of a world which doesn't always take kindly to improvements, corrections or new perceptions.
- Lyelle Palmer, Professor of Education at Winona State College in Minnesota, for his warmth, wisdom, expertise, and wide knowledge of advances in accelerated learning and human development.
- Professor Emeritus Virgil S. Ward, whose vigorous intellectual example kept me from going down for the third and final time at the University of Virginia, and the late

Professor John Curtis Gowan—both of them the world's leading experts on educating the gifted, and both my wonderful friends even though they were not friends of each other.

- Jon Pearson, whose work in "moodgie art" in and around Los Angeles with marginally non- and pre-verbal children has added so much, not only to their lives but to our own theory model in the present work.
- Bill Webber, Cynthia Becker, and Randy Kraft of St. Louis, my valued collaborators on the next generation of Project Renaissance breakthroughs.
- Steve Goldstone of New York City, trainer and practitioner both in this program and in neurolinguistic programming (NLP), whose perspective and experience have helped to enrich both fields.
- Bud Brooksieker, who claims I taught him, but who has taught me far more.
- Lena Borjeson, my publisher and hostess in Sweden, who has so enriched me with her European perspective.
- S.S. Bath, my publisher and host in Singapore and Malaysia, who has immeasurably enriched this program with his astute insight and with the wisdom of the East.
- Professor Luiz Jose Machado de Andrade of State University of Rio de Janeiro, who not only got us published in Brazil but is one of the world's foremost experts and scholars on the limbic brain, is founder and host of one of the great international professional membership organizations in the field of accelerated learning, is creator of the learning theory and method of Emotology, and is a most wonderful friend.
- Octavio Gordillo Gullen, Manuel Garibay Olioquiegui, Rosa Martha, Gloria Gonzoro, and all the members of the CAP Group, my publisher and incomparably gracious host in Mexico City and throughout much of that beautiful country.
- Last, but first and best, I must acknowledge my co-author Richard Poe, not only for heroically working through so much of my dense and turgid prose, but for the initiative

he took which led to this present publication. I must also acknowledge his warmly positive, sensitive, and above all painstakingly accurate handling of reports of this work in his former role as Senior Editor at *Success* Magazine, when first we encountered one another. The lengths to which both he and *Success* went to ensure accuracy were a model of journalistic professionalism, and no doubt account for the wide success of that first-rate publication.

For every one I have mentioned, there are dozens more whom space does not permit me to name, for which I ask their indulgence.

Win Wenger

THE EINSTEIN FACTOR

CHAPTER 1

ARE YOU A GENIUS?

During my twenty-five years in the field of accelerative learning, I have seen the human mind perform many wonders. One such marvel made a deeper impression upon me than most. In 1981 during a seminar I conducted in Ravenna, Ohio, one participant, whom I will call Bob S., had a remarkable—and perhaps lifesaving—encounter with his subconscious mind.

We were practicing a technique called Image Streaming. In this case, I had instructed the group to pair up and take turns describing aloud to their partners, with eyes closed, whatever mental images popped into their heads.

During an Image Streaming session, it is vital to report every image you see, no matter how vague, trivial, or puzzling. But Bob S. had trouble following the instructions. When he closed his eyes, he immediately got a perfectly clear image of an old automobile tire. But instead of reporting it to his partner, per the instructions, Bob tried to block the tire out of his mind. He refused to believe that this was what he was "supposed" to be seeing.

"I kept telling my partner I didn't see anything yet," Bob wrote later, "and kept on trying for something else. But my partner kept after me to describe anything, blobs, lines or whatever. So I finally started yattering about this old tire that kept coming up."

As Bob described the tire to his partner, a realization crept over him. He'd seen this tire before. Indeed, it was the right rear tire of his fiancée's car. But why was it appearing to him now, so vividly and persistently?

"I had the impression there was something wrong with it," Bob recalls. "I excused myself from the exercise, dashed out to a phone and phoned my fiancée. I got her father, and he was the one who went out to check that tire. He found the side of that tire was bruised and cut almost through."

Had the weakened tire blown out on the freeway at 65 miles per hour, it could easily have killed everyone in the car. This message from Bob's unconscious may well have saved his fiancée from deadly peril.

THE SQUELCHER

The incident involving Bob S. stands out not because it was unusual, but because it was typical. Our subconscious minds spew forth streams of images, hunches, and subtle perceptions almost 24 hours a day, many of them charged with insight and premonition every bit as vital as Bob's. Like Bob, most of us fail to heed these messages. Confronted with an urgent, life-saving warning, Bob's first impulse was to *squelch* it. So it is with most of us. Day after day, year after year, the vast majority of people squelch their most profound insights without even knowing it. More than any other factor, this defensive reflex—which I call The Squelcher—blocks us from achieving our full mental capacity.

The Attention Bottleneck

Like most people, Bob had probably been trained from infancy to ignore the stream of perceptions that welled up ceaselessly from his unconscious mind.

"Stop daydreaming!" his teachers probably chided him in school. "Sit up and pay attention!"

Unfortunately, paying attention is a skill with limited usefulness. Scientists calculate that the human brain can pay attention to only about 126 bits of information per second. Simply listening to another person talk occupies about 40 bits of "attention."[1] That leaves only 86 bits to watch the person's facial expression and to think about what you're going to say next (see Figure 1.1).

Yet, our minds are flooded each second by perceptions involving hundreds of times more than 126 bits. Experiments have shown that the human retina can detect a single photon at a time and that the nose reacts to as little as one molecule of scent. These minute perceptions flow constantly into our brains, but they are squelched before they ever reach consciousness. That explains why, in certain rare cases, brain injuries will trigger extraordinary leaps in sensory ability. By a perverse coincidence, such lesions short-circuit The Squelcher and allow in more perception. A neurochemical imbalance called Addison's disease, for example, has been known to heighten the sense of taste by as much as 150 times.[2]

What happens to subtle perceptions when they are squelched? Contrary to common sense, they are neither lost nor destroyed. In fact, the latest evidence suggests that human memory approaches 100 percent retention. We remember potentially *everything*. Yet most of those memories lie so deep in the unconscious that, until quite recently, psychologists had no means of retrieving them other than inducing a profound hypnotic trance.

Figure 1.1 Most people can pay conscious attention to only 126 bits of information per second. Listening to another person talk takes up 40 bits per second, leaving only 86 bits to process other sensory input, such as looking at the person's face. Yet our minds receive far more unconscious information than can possibly squeeze through the Attention Bottleneck.

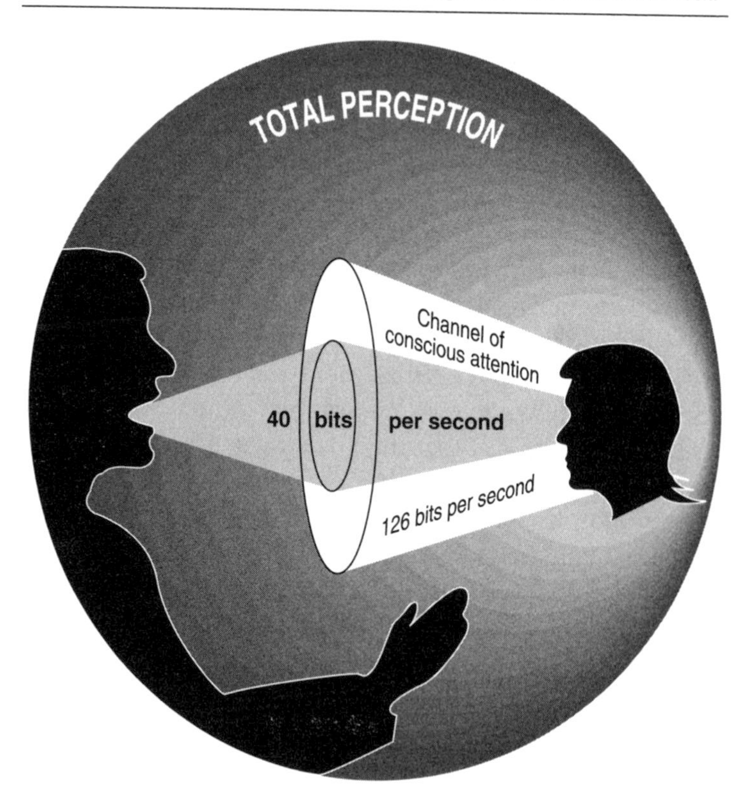

THE GENIUS WITHIN

If you play tennis, your coach has probably told you a hundred times to "keep your eye on the ball." Most of us assume that this means we should "pay attention" to the ball, but that is physically impossible. A tennis ball in flight will always outrun the speed of your conscious thought by about half a second, because it takes a tenth of a second for the image from your eye to reach your brain and another 400 milliseconds for you to form a conscious perception of the ball. If tennis players relied on paying attention, every ball would smack into the fence before they could move their rackets.[3]

Striking a tennis ball is only one of the simpler tasks your unconscious mind can perform. Its capabilities are stupendous. Bob S.'s unconscious took note of a tiny mark on an automobile tire, probably out of the corner of his eye, while Bob was focused on something else. From that wisp of information, it diagnosed a dangerous flaw in the rubber and purposefully strove to bring that problem to Bob's conscious attention—a set of actions that required not only intelligence but a keen sense of responsibility as well.

We might almost imagine Bob's unconscious as a separate person with superior perceptions, constantly looking after Bob from within—an impression strongly reminiscent of the ancient belief in guardian spirits. The Greeks called such supernatural guardians *daemons*. Even the ultrarationalist Socrates credited his daemon with saving his life when he fought in Athens' war with Sparta. The Romans referred to these friendly phantoms as *genii* (*genius* in the singular). It is to such "ingenious" spirits that the ancients attributed all wisdom, insight, and artistic inspiration.

Super Mind

This ancient view is not far from the truth. Each of us does indeed possess a thinking machine vastly superior to our feeble conscious minds. The mathematician John von Neumann once calculated that the human brain can store up to 280 quintillion—that's 280,000,000,000,000,000,000—bits of memory.[4] Many call that a conservative figure.

Estimates of the brain's speed of operation have ranged from 100 to 100,000 teraflops (a teraflop is 1 trillion *flops*, the standard measure of computing speed). Compare that speed to the world's fastest supercomputer, the CM-5, which clanks along at an arthritic 100 gigaflops or 100 billion flops. That's 10^{17} brain flops versus 10^{11} CM-5 flops.

Despite all this awesome computing power in our heads, most of us are hard put to multiply two-digit figures without resorting to a calculator, while even fewer can manage the *New York Times* crossword puzzle or remember what they had for dinner last Wednesday. Only the Mozarts, the Einsteins, and the da Vincis—an infinitesimal sliver of humanity—seem to use their brainpower efficiently (and the evidence shows that even they employ but a fraction of their intellect). So stupendous do their talents seem to the rest of us that we look upon such geniuses much as the ancients did—as divinely gifted beings endowed with what appear to be supernatural powers.

The Elusive Genius

But are geniuses really so different from the rest of us? You would hardly think it by looking at their school records or job histories. Seldom do bona fide geniuses distinguish themselves early in life. Many are labeled "difficult," "slow," or even "stupid." The renowned mathematician Henri Poincaré did so poorly on the Binet IQ test that he was judged an "imbecile."[5] Thomas Edison, whose record 1,093 patents outstripped every inventor in history and transformed human life, was notoriously slow in school.[6]

"My father thought I was stupid," Edison later recalled, "and I almost decided I must be a dunce."[7]

As a child, Albert Einstein, too, appeared deficient to his elders, partly due to his dyslexia, which caused him great difficulty in speech and reading.

"Normal childhood development proceeded slowly," recalled his sister, Maja Winteler-Einstein, "and he had such difficulty with language that those around him feared that he would never learn to speak. . . . Every sentence he uttered, no matter how routine, he repeated to himself softly, moving his lips. This habit persisted into his seventh year."[8]

Young Einstein's poor language skills provoked his Greek teacher to tell him, "You will never amount to anything."[9] Einstein was later expelled from high school and flunked his college entrance exam. After finally completing his bachelor's degree, he failed utterly to attain either an academic appointment or a recommendation from his professors. Forced to accept a lowly job in the Swiss patent office, Einstein in his mid-twenties seemed destined for a life of mediocrity.

But in his twenty-sixth year, Einstein did the unexpected. He published his Special Theory of Relativity—which contained his famous formula, $E = mc^2$—in the summer of 1905. Sixteen years later, he had won a Nobel prize and become an international celebrity. Even today, forty years after his death, Einstein's numinous eyes, bushy mustache, and shock of silver hair remain the quintessential image of "genius," his name a synonym for supernormal intelligence.

WHAT DID EINSTEIN HAVE THAT WE DON'T?

That's what Dr. Thomas Harvey wanted to know. He was the pathologist on duty at Princeton Hospital when Einstein died in 1955. By sheer chance, fate had fingered Harvey to perform Einstein's autopsy. Without permission from the family, Harvey took it upon himself to remove and keep Einstein's brain. For the next forty years, Harvey stored the brain in jars of formaldehyde, studying it slice by slice under the microscope and dispensing small chunks to other researchers upon request. His goal? To uncover the secret of Einstein's genius.

"Nobody had ever found a difference that earmarked a brain as that of a genius," Harvey later told a reporter. ". . . So it was mainly an idea of seeing what we could find."[10]

Harvey himself never found anything, but one of his colleagues did. After examining sections of Einstein's brain in the early 1980s, Marian Diamond, a neuroanatomist at the University of California at Berkeley, announced an amazing discovery—one that may revolutionize ideas about learning and genius.[11,12]

Making a Genius

Most people assume that geniuses are born, not made. But Diamond has devoted her career to creating genius in the laboratory.

In one famous experiment, Diamond placed rats in a super-stimulating environment, complete with swings, ladders, treadmills, and toys of all kinds. Other rats were confined to bare cages. Those rats who lived in the high-stimulus environment not only lived to the surprising age of three (the rat equivalent of 90 in a human), but their brains increased in size, sprouting forests of new connections between nerve cells in the form of dendrites and axons—spindly, branch-like structures that transmit electrical signals from one nerve cell (or neuron) to another. The rats who lived in bare cages stagnated and died younger. Their brains had fewer cellular connections.[13,14]

As long ago as 1911, Santiago Ramon y Cajal, the father of neuroanatomy, had found that the number of interconnections between neurons (called synapses) was the real measure of genius, far more crucial in determining brainpower than the sheer number of neurons.[15] Diamond's experiment showed that—at least in rats—the physical mechanism of genius could be created through mental exercise.

Did this principle apply to people? Diamond wanted to find out. She obtained sections of Einstein's brain and examined them. As she expected, Diamond found an increased number of glial cells in Einstein's left parietal lobe, a kind of neurological switching station that Diamond described as an "association area for other association areas in the brain."[16] Glial cells act as a glue holding the other nerve cells together and also help transfer electrochemical signals between neurons. Diamond expected them, because she had also found high concentrations of glial cells in the brains of her enriched rats. Their presence in Einstein's brain suggested that a similar enrichment process was at work.

Unlike neurons—which do not reproduce after birth—glial cells, axons, and dendrites can increase in number throughout life, depending on how you use your brain. Diamond's work suggested that the more we learn, the more such connections are formed (see Figure 1.2). Likewise, when we cease learning and our minds stagnate, these connections shrivel and dwindle away.

The implication for educators is clear. If Einstein's brain worked anything like the brains of Diamond's rats, it may be possible to create new Einsteins by providing sufficiently stimulating mental exercise.

EINSTEIN'S THEORY OF GENIUS

What sort of mental exercise would correspond in a human to the swings, ladders, treadmills and toys of Diamond's super rats? Einstein himself had some ideas on this subject. He believed that you could stimulate ingenious thought by allowing your imagination to float freely, unrestrained by conventional inhibitions.

For example, Einstein attributed his discovery of the Theory of Relativity not to any special gift but rather to what he called his "retarded" development.

Figure 1.2 Neurons stop reproducing after infancy. But axons, dendrites, and glial cells—which provide electrochemical connections between neurons—keep growing as long as we keep learning. These interconnections are far more important to intelligence than the number of neurons in our brain.

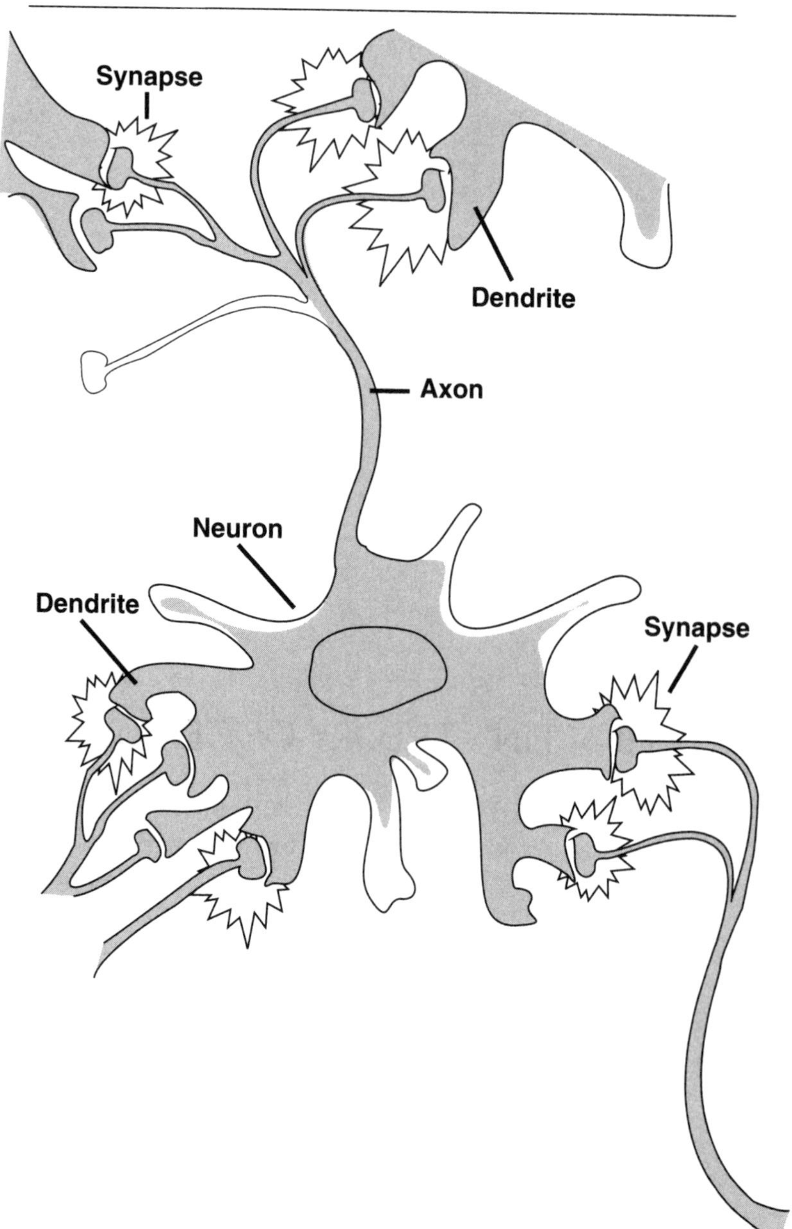

"A normal adult never stops to think about problems of space and time," Einstein mused. "These are things which he has thought of as a child. But my intellectual development was retarded, as a result of which I began to wonder about space and time only when I had already grown up."

Einstein's Ride on a Light Beam

In his last *Autobiographical Note*, Einstein recalled the first crucial insight that led to his Special Theory of Relativity. It came to him unexpectedly, while he was daydreaming, at the age of sixteen.

"What would it be like," Einstein wondered, "to run beside a light beam, at the speed of light?"[17]

Normal adults, according to Einstein, would squelch such a question before it was ever formed in their minds or, having formed it, would quickly forget it. Perhaps that is what Winston Churchill meant when he said that "most men stumble over great discoveries. But most then pick themselves up and walk away."

Einstein was different. With no clear idea of where it would take him, Einstein played with this question for a full ten years. The more he pondered it, the more questions arose. And with each new question, he came closer to the truth.

The "Feel" of Relativity

Suppose, Einstein asked himself some years after posing his original question, that you were riding on the end of a light beam and you held a mirror before your face. Would you see your reflection in the mirror, or not? According to classical physics, the answer was obvious. You would not see yourself in the mirror, because the light leaving your face would have to travel faster than light in order to reach the mirror.

But Einstein could not accept this answer. It fit all the known facts, but for reasons he could not put into words it didn't *feel* right to Einstein. It seemed to him ludicrous that a person would look into a mirror and see nothing there. Trusting his intuition more than he trusted the known and accepted laws of physics, Einstein boldly envisioned a universe that would allow you to see your reflection in a mirror even while riding a light beam. Only years later did he follow up his vision by proving his theory mathematically. It was gut feeling, far more than mathematical rigor, that led Einstein to the answer.[18]

"Invention is not the product of logical thought," Einstein concluded, "even though the final product is tied to a logical structure."[19]

EINSTEINIAN METHOD

With few exceptions, the great discoveries in science have been made through just such intuitive thought experiments as this. Einstein did not invent this technique, but because he was its most famous and active partisan, I have called it Einsteinian Discovery Technique. Another source of Einsteinian Discovery procedures (besides this one) is Sidney J. Parnes' *VISIONIZING: State-of-the-Art Processes for Encouraging Innovative Excellence*, from the Creative Education Foundation.

Psychologist Robert B. Dilts recently studied every surviving scrap of information about Einstein's scientific thought process, drawing upon Einstein's correspondence with Sigmund Freud and mathematician Jacques Hadamard, as well as upon detailed interviews that Einstein granted to psychologist Max Wertheimer, the founder of Gestalt therapy. Dilts's biographical study yielded some remarkable insights.

"Instead of words or mathematical formulas," Dilts concluded in his three-volume *Strategies of Genius* (Meta

Publications, Capitola, CA, 1994), "Einstein claimed to think primarily in terms of visual images and feelings. . . . Verbal and mathematical representation of his thoughts came only *after* the important creative thinking was done."[20]

Combinatory Play

In fact, Einstein attributed his scientific prowess to what he called a "vague play" with "signs," "images," and other elements, both "visual" and "muscular."

"This combinatory play," Einstein wrote, "seems to be the essential feature in productive thought."[21] (See Figure 1.3.)

Regarding his Theory of Relativity, Einstein told Max Wertheimer, "These thoughts did not come in any verbal formulation. I very rarely think in words at all."[22]

In Einstein's vague play with images and feelings, I see a mechanism at work similar to that which helped Bob S. save his fiancée's life. Einstein and Bob S. both lacked the capacity to solve their respective problems through conscious thought. Bob S. achieved his insight through the technique of Image Streaming. Einstein used a private method of his own. But both drew upon subtle insights from outside the conscious realm.

THE KNACK OF GENIUS

Over the years, my studies have led me consistently to the conclusion that geniuses are little more than ordinary people who have stumbled upon some knack or technique for widening their channel of attention, thus making conscious their subtle, unconscious perceptions.

They usually develop this knack so early in life that they completely forget the secret by the time they grow up. It becomes automatic. Most geniuses are therefore every bit

Figure 1.3 Einstein ascribed his ingenious insight to a "combinatory play" of sense impressions, "muscular" feelings, emotion, and intuition. Only in the final stages of thought did Einstein translate his theories into words and equations.

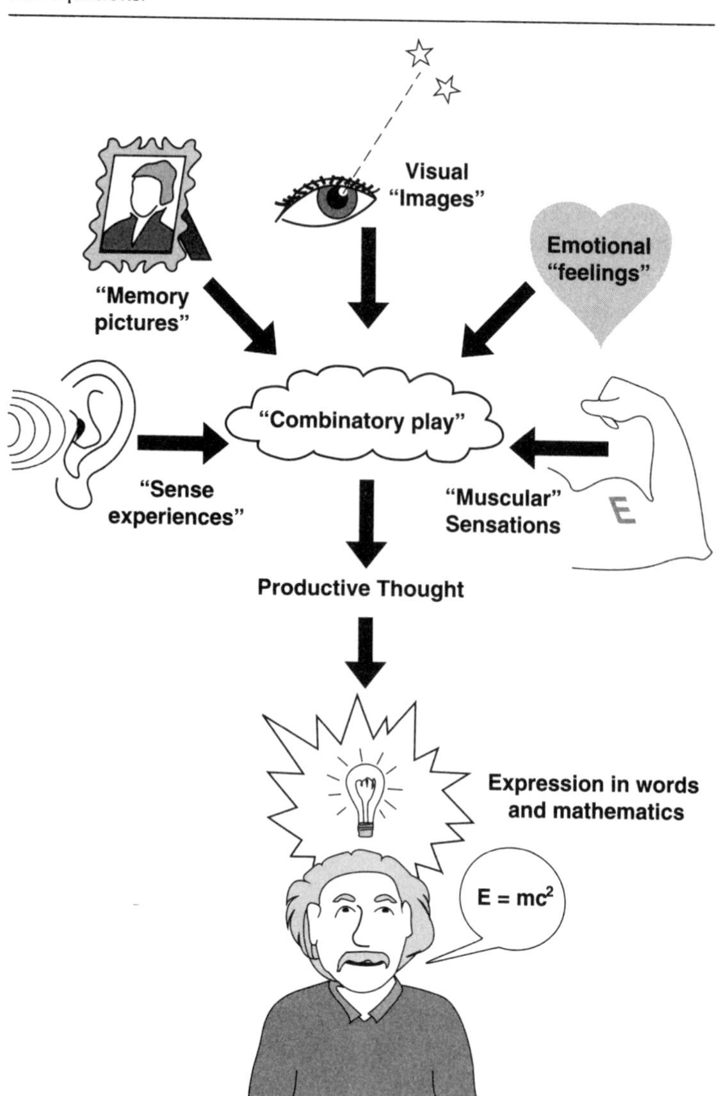

as mystified as the rest of us as to how they manage to achieve such extraordinary results.

A Baseball Genius

Some years ago, I visited a friend in Chicago. My friend's son was trying out for the high school baseball team but feared he wouldn't make the cut because of his poor batting average. I worked with the boy for about an hour, employing many of the techniques that you will learn to use later in this book.

In the course of our session, the boy discovered that he had the greatest success when he imagined a tiny flyspeck on the baseball and aimed his bat at that flyspeck rather than at the ball itself. This flyspeck gave him just the extra focus he needed to connect with the ball.

It may seem a trivial insight, but its effect on the boy's game was astonishing. In baseball, a .250 to .300 batting average is considered quite good. But during the first ten games of the season, this boy batted .800! He not only made the team but went on to be named Most Valuable Player for both the team and the league for that year.

In a single one-hour session, we had succeeded in identifying a technique that made this boy a baseball genius. But the most surprising discovery was yet to come. I did not see this boy again until several years later. He was still playing baseball, and he clearly remembered our one-hour session as having marked the turning point in his athletic career.

But the boy had entirely forgotten the details of what he had learned during our session. He remembered nothing about the flyspeck and no longer consciously envisioned it when striking the ball. Indeed, he was as mystified as his teammates as to just how he had become such a great batter so quickly.

The Talent Question

It's easy to argue that this boy must have had a talent for baseball all along. I'm sure he did, but when I met him there was no talent in evidence. By every objective standard, he was doomed to fail. Only when he discovered the trick of the flyspeck were his latent talents catalyzed.

In fact, all of us possess hidden talents, often in the very areas where we think ourselves least capable. Study, practice, and hard work can bring about incremental improvement. But if we wish to unleash the full power of our genius, we must find that crucial catalyst, that simple trick or knack that will bring our bodies, senses, and minds into critical focus.

I call that catalyst the Einstein Factor.

YOU ARE BRIGHTER THAN YOU THINK!

Conventional education and job training are notoriously effective at crushing our confidence and squelching our most brilliant thoughts. Most of us learn early to suppress our natural genius. Like young Thomas Edison, we allow others to beat us down, until we see only a dunce when we look in the mirror.

In reality, you are far brighter than you think. The techniques in this book will help you reverse years of conditioning. They will help you find the "flyspeck" that will work for you. No two people access the Einstein Factor in exactly the same way. I can't predict what form your flyspeck will take, but your unconscious mind can. Like Aladdin's jinn, your unconscious mind can emerge as a powerful and everpresent ally, if you only allow it.

CHAPTER 2

YOU ARE ALWAYS DREAMING

Inventor Elias Howe had labored long and hard trying to invent the first sewing machine, but nothing worked. Then, one night, Howe had a horrible nightmare. He was running from a band of cannibals. They were so close he could see the gleam of their spear tips. Through his terror, Howe suddenly noticed that each spear point had a hole bored in its tip shaped exactly like the eye of a sewing needle. At that moment, he awoke, gasping with fear.

Only then did Howe realize what his nightmare was trying to tell him. To make his sewing machine work, he would have to move the eyehole from the middle of the needle down to the tip. It was precisely the breakthrough he had been seeking. Thanks to Howe's dream, the sewing machine was born.

THE PROBLEM WITH DREAMS

History abounds with stories like Howe's. Dreams have inspired rulers, artists, scientists, and inventors since biblical times. But they have their limitations. Dreams are notoriously hard to control. We have not yet learned how to summon them at will (although someday we may). We cannot decide in advance what we will dream about. And, most of the time, we forget our dreams anyway.

Lucid Dreaming

Psychophysiologist Stephen LaBerge has tried to solve this problem through his technique of "lucid dreaming." He trains people to become fully conscious while in the midst of a dream.

The first step is to get into the habit of asking yourself "Is this a dream?" at frequent intervals throughout the day. Once the habit becomes ingrained, you will eventually remember to ask yourself that same question while you are dreaming. The first time you ask this question and the answer comes up yes, congratulations! You will have begun your first lucid dream.

Lucid dreamers soon learn to control the action in their dreams and to choose the experiences they want. Many report exhilarating episodes of soaring through the clouds, exploring underwater realms, changing into animals, and performing other amazing feats at will. They converse at length with interesting dream characters, work out neuroses, and confront fears. They can even stop a nightmare in its tracks and steer it toward a more pleasant conclusion. Interested readers should investigate LaBerge's remarkable books, *Lucid Dreaming* (1985) and *Exploring the World of Lucid Dreaming* (1990).

LaBerge's method holds great promise as a tool for unlocking unconscious riches, but it is not the panacea some enthusiasts claim. Critics of lucid dreaming say LaBerge downplays the difficulty of learning the technique. It seems to come naturally to some people and much harder to others. Often it requires equipment that can cost from $275 to $1,000, such as the NovaDreamer, a device that cues the user with lights or sounds at the moment he enters REM sleep, the stage of Rapid Eye Movement when dreaming typically commences.

Even if these difficulties could be surmounted, it has been my experience that dream imagery is irrelevant and misleading as often as it is useful. The profound trance of REM sleep opens the mind to deep-rooted thoughts and emotions so far removed from our everyday concerns that they are often impenetrable. There is always great benefit to be gained from dreamwork, and it never ceases to fascinate, but if your aim is to enhance creativity and solve specific problems, you should supplement your arsenal with far more direct and efficient methods.

YOU CAN "DREAM" WHILE YOU'RE AWAKE

How, then, can you best gain access to the remarkable flow of unconscious perception? Over the last twenty-five years, I believe I have found an answer. The Image Streaming technique I developed opens the mind to a flow of symbolic imagery as potent as that of any dream. But, unlike dreaming, you can practice Image Streaming while you're wide awake, and you can do it virtually anytime, anywhere. Ten minutes of Image Streaming every day will suffice to induce profound, positive change in your life.

The Image Stream

The fact is that we are *always* dreaming. Psychologists estimate that we spend about 50 percent of our time daydreaming and over 8 percent of it sleep-dreaming. That means we spend 58 percent of our lives absorbed in passive reception of subconscious imagery. It sounds like a lot of time, but in fact these figures are a gross underestimate.[1,2]

Evidence suggests that the Image Stream literally *never* ceases. Even when our minds are preoccupied with work, conversation, or other demanding tasks, the sensory mechanisms of our minds continue to generate imaginary sights, sounds, smells, tastes, and feelings. Many of these images consist of memories that are triggered by random associations. Others are echoes or reinforcements of our conscious thoughts at the moment.

This is one of the few places where The Squelcher makes itself useful. If you did not squelch your Image Stream while flying a plane or performing surgery on a patient, you would endanger people's lives. The flow of images would distract you from doing your job.

A PECULIAR AFFLICTION

The great Yugoslav inventor Nicola Tesla suffered from what he called a "peculiar affliction." As a boy, Tesla was tormented by flashing lights and images of remembered scenes that would appear before his eyes without warning and with blinding intensity. A single word spoken to Tesla in conversation might suddenly trigger a lifelike image of the person or thing that the word represented. These images were so real that Tesla found himself "quite unable to distinguish whether what I saw was tangible or not," a confusion which caused him "great discomfort and anxiety."[3] He eventually overcame this affliction through mental exercise and sheer willpower.

In the 1920s, Soviet psychologists began studying a journalist named Solomon Shereshevesky in an attempt to plumb the secret of his near-perfect memory. Shereshevesky's unusual talent brought with it some crippling problems. Words in conversation, as well as random thoughts and memories, could trigger intense storms of sensory impression running the gamut of all five senses.

"I can't escape from seeing colors when I hear sounds," he recounted. "If, say, a person says something, I see the word; but should another person's voice break in, blurs appear. These creep into the syllables of the words and I can't make out what is being said."[4]

While talking to the great Soviet psychologist L. S. Vygotsky, Shereshevesky remarked, "What a crumbly yellow voice you have." Shereshevesky liked the voice of filmmaker S. M. Eisenstein much better. "Listening to him, it was as though a flame with fibers protruding from it was advancing right toward me. I got so interested in his voice, I couldn't follow what he was saying."[5]

On one occasion, Shereshevesky approached a street vendor and asked what flavor ice cream she had. "Tutti-frutti," she replied.

"But she answered in such a tone," he recounted, "that a whole pile of coals, of black cinders, came bursting out of her mouth, and I couldn't bring myself to buy any ice cream after she answered that way."[6]

Tesla and Shereshevesky were not hallucinating in the clinical sense. Their subconscious minds were reacting to the world in quite normal fashion. These men simply lacked the ability that most of us have to squelch their Image Streams when necessary.

Tesla's Gift

While troublesome, the intensity of Tesla's Image Stream appeared to stimulate his genius. Among his many talents, Tesla possessed the remarkable ability to visualize his inventions in minute detail before even beginning to write them down. He would mentally build a new device part by part and test-run it, all in his imagination. So accurate were Tesla's mental blueprints that he could diagnose a problem with a machine by the way it ran in his mind.

"It is absolutely immaterial to me whether I run my turbine in thought or test it in my shop," he wrote. "I even note if it is out of balance. There is no difference whatever, the results were the same."[7]

By this means, Tesla developed all the basic mechanisms of today's global electric power grid, including high-voltage transformers, long-distance transmission lines, hydroelectric generators, and alternating current.

The Man Who Remembered Everything

Shereshevesky, too, put his image stream to work in performing remarkable feats of memory. By the time he died in 1950, Shereshevesky had become world-famous as "the man who remembers everything." His memory was literally perfect. On one occasion, psychologist A. R. Luria gave him a long list of nonsense syllables that began like this:

1.	ma	va	na	sa	na	va
2.	na	sa	na	ma	va	
3.	sa	na	ma	va	na	
4.	va	sa	na	va	na	ma
5.	na	va	na	va	sa	ma
6.	na	ma	sa	ma	va	na
7.	sa	ma	sa	va	na	
8.	na	sa	ma	va	ma	na

Because the list was so long and the syllables so meaningless and similar in form, ordinary mnemonic tricks would offer no help in trying to recall them. Shereshevesky not only memorized the list easily, but when Luria popped a surprise quiz on him *eight years later,* Shereshevesky reeled off every syllable without a single mistake, in perfect order and in their proper columns.

SHERESHEVESKY'S TRICK

Shereshevesky's trick was to apply his multisensory thoughtstorms to jog his memory. He claimed to memorize words not only through photographic mental images but also by their "taste or weight" and "a whole complex of feelings."[8]

For instance, when he was first read the above list of nonsense syllables, Shereshevesky suddenly imagined himself in a forest. A thin, grayish yellow line appeared to his left.

"This had to do with the fact that all the consonants in the series were coupled with the letter *a*," he later explained. "Then lumps, splashes, blurs, bunches, all of different colors, weights, thicknesses rapidly appeared on the line. These represented the letters *m*, *v*, *n*, etc."

To retrieve the syllables later, Shereshevesky explained, he would simply walk down the same imaginary path in the woods, "to grope at, smell, and feel each spot, each splash."[9]

Sorcerer's Apprentice

Shereshevesky's incredible memory carried a price. Like the spell cast by the sorcerer's apprentice, it ran rampant in unpredictable ways. For one thing, he found it almost impossible to forget. This caused unexpected problems for him. If, for example, he tried to memorize lists written on a blackboard, his memory might get tangled up with other lists written on that same blackboard at other times. Shereshevesky had to struggle constantly to squelch certain memories in order to make room for others—a process that comes naturally to most people.

He also had trouble with faces. Most of us remember a familiar face only as an overall pattern. Not Shereshevesky. When he tried to recall a face, his mind would swarm with images of that face in every detail, from every perspective, in every kind of lighting, and with every different nuance of expression it had ever worn. In all this confusion, he could seldom recognize whose face it was.

"They're so changeable," he complained. ". . . It's the different shades of expression that confuse me."[10]

STRIKE A BALANCE

Genius does seem to be linked to the intensity of our subconscious imagery, but to be effective we must strike a balance between Squelcher and Image Stream. In striving to gain access on demand to intense and vivid imagery, we must also preserve the ability to squelch it at appropriate times. This balance is best achieved through a controlled process like Image Streaming, which allows us to choose the time and place of our imaging and to remain completely conscious and alert throughout the session.

THE DREAM RECALL PHENOMENON

We have far more ability than we commonly realize to engineer our consciousness. The phenomenon of dream recall offers one of the clearest illustrations of this ability. Some experts, such as neuroscientist Allan Hobson, maintain that when our brains enter REM sleep they stop emitting the neurotransmitter chemicals necessary for memory. It thus becomes physiologically impossible for us to remember most of our dreams, Hobson says.[11] Yet you can easily demonstrate to yourself the fallacy in this theory.

Try a simple experiment. Ask yourself "What were the plots of my first three dreams last night?" If you're like most people, you will not be able to recall a single dream from last night, much less three of them in a row. Indeed, many people will claim that they do not dream at all, except on rare occasions. But sleep researchers assure us that the average person has about five separate dreams each night, lasting a total of at least two hours. We simply forget them by the time we awake.

Now try this. Before you go to sleep tonight, leave a notebook and pen by your bedside. The instant you wake up—even if it is in the middle of the night—write down everything you remember about your dreams. You may not have any results for the first few days, but I guarantee that after a couple of weeks of this exercise you will find yourself waking up each morning with vivid, detailed memories of three or more different dreams. You will remember such an astonishing mass of material that you will soon despair of writing it all down. Doing this exercise with a tape recorder will produce the same results.

You Get More of What You Reinforce

This well-known phenomenon illustrates what we might call the First Law of Behavioral Psychology: Whatever you reinforce, you will get more of.

Every time you write down your dreams, you reinforce the behavior of dream recall. Your recollection thus becomes stronger. Likewise, every time you *fail* to write down a dream, you reinforce the behavior of forgetting your dreams, and your recall correspondingly weakens.

Image Streaming works the same way. As you begin this book, your Image Stream is probably very weak, because you have spent your life ignoring and suppressing it. But as you begin daily Image Streaming, you will be startled to find the imagery growing stronger and more vivid as your practice progresses.

HOW TO IMAGE STREAM

The procedure of Image Streaming is deceptively simple. You sit back in a comfortable chair, close your eyes, and describe aloud the flow of mental images through your mind. Three factors are absolutely crucial. I call them the Three Commandments of Image Streaming:

1. You must describe the images *aloud*, either to another person or to a tape recorder. Describing them silently will defeat the purpose of the exercise.

2. You must use *all five senses* in your descriptions. If you see a snow-covered mountain, for example, don't just describe how it looks. Describe its taste, its texture, its smell, and the sound of wind howling across its peak.
3. Phrase all your descriptions in the *present tense*.

THE REINERT EXPERIMENT

I originally developed Image Streaming as a kind of oracle on demand. The idea was that you would pose a question to your subconscious mind and receive an answer through mental imagery, much as Elias Howe received help on his sewing machine from the dream world. In fact, Image Streaming has proved extremely effective for this purpose, as I will describe in later chapters.

But Image Streaming also revealed unexpected benefits. One of its more surprising side effects came to light in an experiment performed by Dr. Charles P. Reinert, a physics professor at Southwest State University in Marshall, Minnesota. Reinert asked seventy-nine of his first-year students to take part in a test of accelerated learning techniques. During the 1988 winter quarter, some students agreed to practice the Whimbey Method, a standard program that uses word problems to build analytical skills. Others were to try an unusual new method called Image Streaming.

Reinert administered a standard intelligence test to each student before and after the course. The results proved startling (see Figure 2.1). In one section, students who had practiced the Whimbey Method gained the equivalent of about 0.4 IQ points per hour of practice. But those who Image Streamed gained a whopping 0.9 points per hour—the equivalent of a full IQ point for every 80 minutes of practice![12,13] (Reinert's results are not yet published and should be considered preliminary. Definitive figures will

Figure 2.1 A physics class at Southwest State University in Marshall, Minnesota, was divided into two groups, one practicing Image Streaming throughout the course and the other practicing the Whimbey Method, a standard program for building analytical skills. Intelligence was measured at the beginning and end of the course. The Image Streamers showed much sharper gains in IQ.

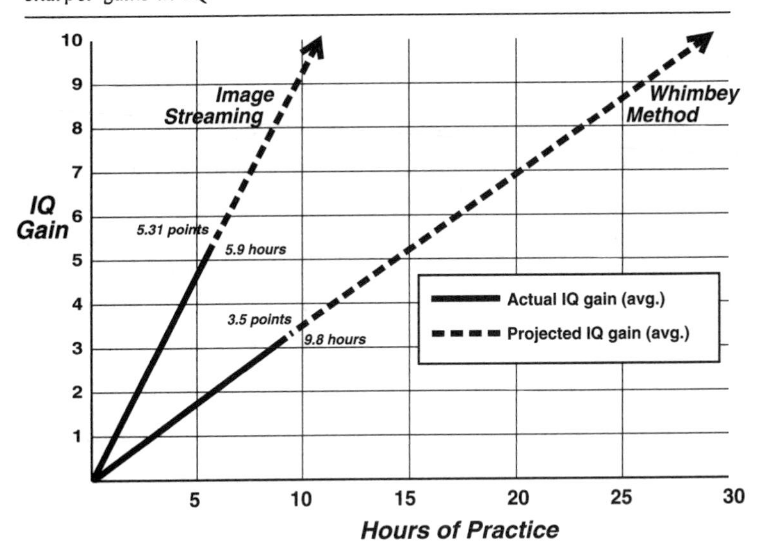

be published only after Reinert completes a long-term study now in progress.)

Digging Channels in the Brain

Why would Image Streaming raise IQ? The answer lies in our brain cells. Consider what happens when we practice a new skill, such as riding a bicycle. At first, it seems impossible to balance atop a two-wheeled vehicle on the move, but once we have mastered the skill we never forget it. We can come back after twenty years of little or no practice and still remember how to ride a bike.

That's because the skill of bicycle riding was imprinted on our brains in the form of a vast, intricate network spanning millions of different neurons. Such networks can be huge. Scientists at New York University trained a cat to distinguish between two doors, one with a pattern of two concentric circles on it. The cat easily learned that the door with the circles could be pushed open to reveal a bowl of cat food. But the experimenters were shocked when a radioactive brain scan revealed that this simple act of recognition and discrimination lit up 5 to 100 million of the cat's neurons at a time—one-tenth of its entire brain mass![14]

This experiment showed that the same neurons must be used for many different memory networks simultaneously. Otherwise, there would be room for only ten complex tasks in the cat's entire brain. Memories are stored not in the cells themselves but in the overall pattern of electrical signals firing *between* cells. How are these patterns preserved? If you stop riding a bike for twenty years, does your brain keep firing an electrical bike-riding pattern nonstop for the next two decades? Of course not. Your brain would quickly burn out if it had to keep all 280 quintillion bits of your memory—including every skill you ever learned—lit up permanently and simultaneously.

Psychologist Donald Hebb discovered in the 1940s that when two adjacent neurons get into the habit of firing signals to each other, neurochemical changes take place in both cells that make it easier for them to interact with each other than with other neurons that are *not* involved in the same learned activity. When you learn to ride a bike, those neurons involved in the distinctive pattern of bike riding form long-lasting Hebbian connections. When you mount a bike after twenty years of nonriding, those connections are still present, and the electrical impulses flow through them much as rainwater tends to flow through channels that have already been eroded in the earth.[15]

Like bike riding, Image Streaming is a learned skill. The more we practice it, the more deeply we reinforce the millions of Hebbian connections linking certain critical portions of our brain.

SLEEP PARALYSIS

A sixty-seven-year-old man periodically dreamt that he was riding a motorcycle. In his dream, a rival motorcyclist would try to run him off the road, and the man would fight back, trying to kick his attacker.

Unfortunately, his violent dream spilled over into real life. The man's wife complained that he punched and kicked her in his sleep and sometimes thrashed wildly around the room. Psychiatrists at the University of Minnesota concluded that the man suffered a malfunction in his sleep paralysis mechanism.

Key portions of our brain don't distinguish between dream and reality. When we see, touch, taste, smell, or hear things in a dream, the corresponding sensory portions of our brain light up, exactly as if we had sensed those things in real life. Physical movement in dreams likewise activates the corresponding motor mechanisms that would drive such movement in waking life.

"As far as the neurons are concerned," says neuroscientist Allan Hobson, "the brain is both seeing and moving in REM sleep."[16]

To keep us from harming ourselves and others, a safety switch deep in our brain stem largely shuts off our muscular system during REM sleep, effectively paralyzing us. That safety mechanism broke down in the case of the man described above.[17]

POLE-BRIDGING

Just as in a dream, the imaginary sights, sounds, and sensations of Image Streaming activate the appropriate brain centers in a near-perfect simulation of reality. Neurologically speaking, an Image Streamer talks, listens, sees, smells, tastes, feels, analyzes, reflects, wonders, creates, and generates mental imagery *all at the same time.*

This unusual combination of mental activities spans or bridges many opposite "poles" of the brain (see Figure 2.2). Over the last fifteen years, the quest to achieve balance

Figure 2.2 When you practice Image Streaming, every major "pole" of your brain is brought into simultaneous play. This Pole-Bridging effect exercises the same mental muscles that Einstein used to formulate his Theory of Relativity.

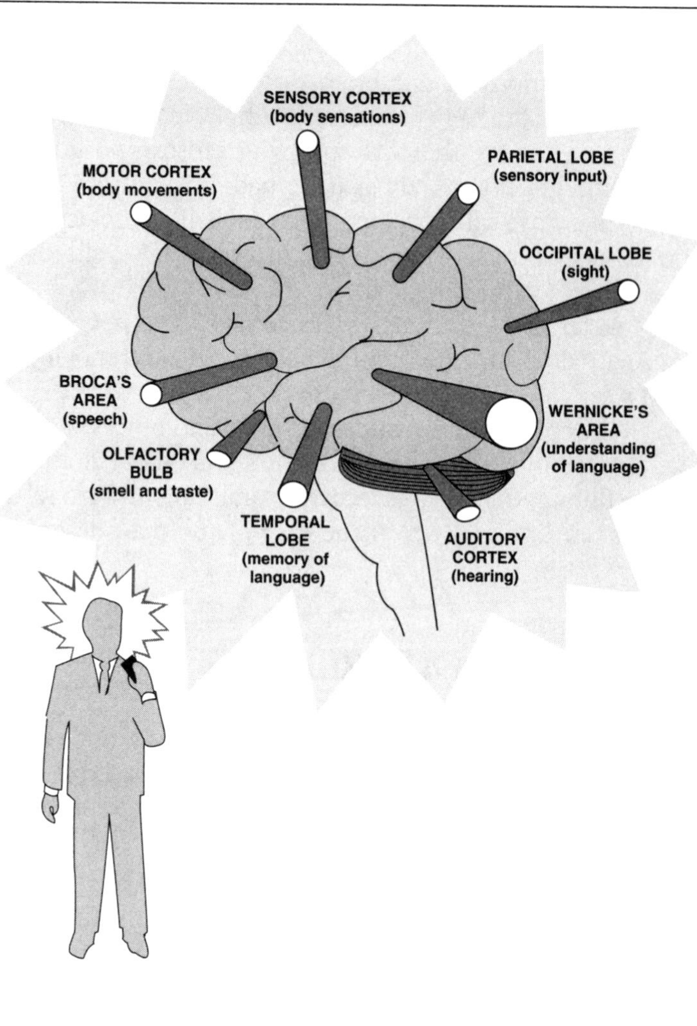

between the brain's analytical left hemisphere and its creative, pattern-sensing right hemisphere has become almost a fad. In the most prestigious corporations, trainers routinely drill stodgy executives in "right-brain" thinking. The more adventurous visit brain salons, where strobe lights and specially modulated sound waves purportedly enhance "whole-brain" synchrony.

But the left-right dichotomy has been somewhat overplayed. In fact, important brain functions are just as likely to be subdivided between the top and bottom or front and back of the brain. Virtually any activity that links opposite sides or "poles" of the brain contributes toward the brain's overall balance.

Of all the methods I know, Pole-Bridging seems to activate the largest and most balanced portion of our brain mass, pushing conscious activity to its utmost limit and building powerful networks of Hebbian connections that crisscross the brain not only from right to left but also from top to bottom and from front to back. Image Streaming is only one of many such Pole-Bridging exercises, but it is arguably the most effective.

Balance in the Brain

The Pole-Bridging effect was given startling confirmation by Dr. Reinert's experiment. In addition to increasing IQ, the Image Streaming technique was found to induce a more balanced learning style in users as measured through a standard test called the Kolb Learning Style Inventory, which was administered to students before and after taking Reinert's physics course.

The Kolb test determines the degree to which a person relies on four different styles of learning: concrete experience, abstract conceptualization, reflective observation, and active experimentation. Ideally, a person should balance all four styles. Students in Reinert's class who used the Whimbey Method veered sharply away from the median, becoming significantly more reflective and concrete but less active and abstract—an effect that occurs during the course of any typical college semester, with or without use of the Whimbey Method, according to some studies. Image Streamers, on the other hand, moved uniformly toward an ideal learning style balanced among all four categories.[18]

LET'S GET STARTED

By now, you're probably eager to start Image Streaming. As with all things in life, some people will prove more naturally adept than others. And as with most endeavors your skills will improve with practice. The next chapter includes a wealth of proven techniques for jump-starting even the most sluggish of Image Streams.

CHAPTER 3

OPEN YOUR IMAGE STREAM

o you're ready to Image Stream. You find a comfortable chair, turn on your tape recorder, sit back, close your eyes and . . . nothing happens! Where are the images?

Alas! You are among the 30 percent of the population who have difficulty generating mental imagery. But don't despair. *Everyone* has an Image Stream. You simply need to learn how to stop squelching yours. Even if you are among that happy 70 percent who can already produce images at will, the techniques in this chapter will help you enhance the power and clarity of your Image Stream.

INCREASE YOUR NEUROLOGICAL CONTACT

Try this experiment: Pick two corners or sections of the room you are in. Now take a piece of paper. On one side of the paper, write a description of the first corner you chose. On the other side of the paper, write a description of the second corner.

When you describe the first corner, confine your description to terms involving color, texture, form, feel, and sense of position in space. When you describe the second, use only abstract terms that have nothing to do with sensory impressions. For example, you might write, "There's a picture hanging on the wall and an upholstered chair wedged in the corner," but nothing about how those objects look or feel. Take about 5 minutes to write out your description (about 3 minutes if you are using a tape recorder).

Now look over your results. Which description is more interesting? Which conveys more of the experience? Which puts the reader more intimately in touch with the corner being described?

Obviously, the first description is more vivid. That's because it gives the reader a greater degree of *neurological contact* with the corner. When you hear or read a description overflowing with vivid sensory impressions, your brain automatically starts to light up in the appropriate sensory areas, just as it lights up in a dream. The more senses you evoke, the wider the base of neurological contact.

A COMMINGLING OF THE SENSES

Walt Disney was a great lover of classical music. He claimed that listening to the masters made pictures in his head. In an attempt to share this experience, Disney created *Fantasia*, an animated film in which classical music springs to life in a phantasmagoria of shapes, sights, and colors.

Disney's multisensory approach was designed to widen the film's neurological contact with its audience. He fervently believed that it would convey the music more powerfully.

"There are things in that music that the general public will not understand until they see things on the screen representing that music," he said. "Then they will *feel* the depth in the music."[1]

Many experts today believe that Disney must have been at least slightly synesthetic. This is a natural condition that occurs in a little less than 1 percent of the population.[2] It can also be induced temporarily with drugs. Early LSD researchers discovered that psychedelic compounds tend to break down the boundaries between different senses. Under their influence, you might "hear" the color red or "smell" a Bach concerto. Evidence is growing that this commingling of the senses—called *synesthesia*—may be a normal function of the mind that is simply suppressed in most people. Image Streaming seems to draw much of its Pole-Bridging power from this hidden mechanism, playing upon links between senses that most of us think of as quite distinct and separate.

A Synesthetic World

Neurologist Richard Cytowic has spent years studying synesthetes, people who are born fully synesthetic. Such people may see golden balls when hearing a vibraphone or a glass column when they taste spearmint. Some feel geometric shapes pressing against their skin on tasting certain foods or even twist their bodies involuntarily into characteristic shapes in response to hearing specific words. This condition brings to mind the splashes, lines, and colors the Russian journalist Shereshevesky saw when certain words were pronounced. Shereshevesky was, in fact, a classic synesthete.

While conducting a radioactive brain scan on one synesthetic subject, Cytowic was shocked to see a wholesale diversion of blood flow from the cerebral cortex as the man entered a synesthetic experience. "We have never, never seen anything like it," Cytowic later remarked.[3] The cortex, or "gray matter" is usually considered the most human part of the brain, responsible for higher intellectual thought. Because blood was diverted from the cortex during synesthesia, Cytowic hypothesized that commingling of the senses must occur deep in the limbic system, the instinctive portion of the brain that gives rise to primitive drives such as hunger, emotion, and sexual desire.

In nonsynesthetic people, the cortex acts as a Squelcher, suppressing synesthesia and keeping it safely corralled in the limbic brain. On a conscious level, most of us therefore perceive sharp boundaries between the senses. But our unconscious minds apparently function in a fully synesthetic world.

Seeing without Seeing

Neurologist Antonio Damasio studied victims of prosopagnosia, a condition in which brain injury has destroyed a person's conscious ability to recognize a face. Damasio showed his subjects photos of different faces—some of friends and family, others of famous people, and still others of perfect strangers. The subjects failed to consciously recognize any of the faces. But when they looked at a familiar face, the electrical conductivity of their skin jumped dramatically—a sign of emotional response. Damasio concluded that his subjects could indeed recognize faces on a subconscious level, but because of their injury this recognition could not penetrate into the conscious mind.[4]

A similar effect occurs during "blindsight." People who have been blinded by brain injury, rather than by damage to the eyes or optic nerves, are apparently still able to see. Their brains simply can't make sense of the visual impulses. When Dr. Anthony Marcel of Cambridge University asked blindsighted patients to retrieve certain objects he had placed in front of them, they reached out and grabbed the objects deftly, without groping or fumbling, in a way that would have been impossible without sight.[5] Other experimenters have shown that blindsighted people can pick out specific shapes from an array of different forms on request.

In a similar way, synesthetic perceptions seem to flood our cortex from the limbic brain without most of us being aware of them. The Squelcher blocks these confusing signals from our consciousness, but their subtle effects nonetheless infuse our experience as unmistakably as the galvanic skin responses of Dr. Damasio's prosopagnosics and the startling dexterity of Dr. Marcel's blindsighted patients.

Such synesthetic vestiges emerge in common turns of speech, such as when we speak of the "coolness" of the color blue, the "sweetness" of a woman's voice, or the "piercing" quality of a sound. These metaphors make no rational sense, yet we understand them instinctively.

"You know why they have music in restaurants?" asked one synesthete. "Because it changes the taste of everything. If you select the right kind of music, everything tastes good. Surely people who work in restaurants know this."[6] Perhaps they do. But if restaurateurs know the reason for their curiously apt choices in music, it is only unconsciously. Thus, the unending stream of unconscious perception enriches and stimulates our lives in hidden ways.

Don't Panic!

Some readers may fear that I'm about to tell them they must become synesthetes. Don't panic! Nothing could be farther from my intention. In fact, full-fledged synesthesia is unusual, unnecessary, and sometimes unpleasantly distracting, as we saw with the case of Shereshevesky. Dwelling upon it consciously can be as futile and enervating as obsessing over our own heartbeat or trying to feel the secretion of our glands. As with so many other bodily functions, synesthesia does its best work in the dark, when we are totally unaware of it.

But its work is critically important to Image Streaming. Remember that the Image Stream makes use of *all five senses*, not just sight. Your brain is wired in such a way that vision will always tend to dominate the creative process. That's all right. But when we describe mental images into a tape recorder, we should take care to include in our descriptions other senses as well, especially taste, scent, and touch (texture), which are most often neglected.

You don't have to hear your own heartbeat to know that jogging helps your cardiovascular system. Likewise, you don't need to be a synesthete to exercise your synesthetic "muscles." Multisensory description builds up the neural connections between your senses and widens your neurological contact with the Image Stream. The effects of such exercise will be seen quickly in stronger, clearer, more solid-looking mental images.

SELF-REINFORCEMENT

The strength and vividness of mental imagery are potentially limitless. Sights, sounds, and sensations, if imagined to their full potential, can solidify so strongly as to be indistinguishable from real perceptions.

Given a highly responsive subject, a skilled hypnotist can induce what is called a "positive hallucination," commanding the person to see and even converse with a companion who isn't really present. The hypnotist can also induce a "negative hallucination," wherein a real, flesh-and-blood companion actually present will suddenly become invisible.

Such trance-induced hallucinations draw their potency from the power of suggestion. Under hypnosis, a less suggestible subject might hallucinate a person only as a transparent, ghostlike wraith. But a profoundly suggestible subject will perceive his imaginary companion as completely solid and as so realistic that, if he reaches out and touches her, he will feel the warmth of her skin.[7]

Image Streaming, on the other hand, does not rely upon a hypnotist's suggestion for its vividness. It is a self-reinforcing process. The more absorbed you become in describing an image, the more intently your mind focuses on it and the clearer, sharper, and more solid it becomes. Your own objective observation, rather than someone else's suggestion, gives life to your Image Stream.

BUILD YOUR POWERS OF DESCRIPTION

Let's get started. The first step is to build your powers of description. Like any other skill, this ability is acquired through practice. The best way to break the ice is to simply start describing something.

Begin with your physical surroundings. Describe the room where you are sitting or some outdoor scene you often pass in the course of your day. It is extremely important that you describe it *aloud* and *to a tape recorder or another person.* I cannot repeat these instructions too many times. Experience has shown that the Image Streaming procedure loses its effectiveness if you leave out any part of it.

Think of the tape recorder as a telephone. Imagine that you are describing the scene to a friend. Your goal should be to describe it so richly that you literally force the reality of it onto your listener through the sheer richness of detail and raw sensory description.

When in doubt, keep talking. Don't edit. Don't worry about speaking nice sentences. If you're wondering whether to include some nuance or some trivium in your description, go ahead and describe it. There is no right way to describe something. The only mistake you can make is to hesitate, to stop, or to edit. These processes are The Squelcher at work in your mind.

After a few days of diligent practice, your ability to describe your surroundings will have vastly improved (and so also will your powers and richness of observation). As soon as you have grown comfortable with the descriptive process, the next step is to start describing scenes and pictures that aren't physically there but exist only in your mind.

DO IT YOURSELF

We are now coming to the nuts and bolts section of this chapter. In the following pages, you will find a number of step-by-step instructions for various techniques. After reading the instructions for each technique, for example, Velvety-Smooth Breathing, *put down the book* and actually *try* the technique before moving on to the next section.

I strongly recommend that as you progress through the book you try out each and every technique at least once, as soon as you encounter it in the text. Your grasp of the material will thus be self-reinforcing rather than abstract.

I will refrain from pestering you further by harping on the importance of practice in succeeding pages. However, I cannot emphasize sufficiently how much richer will be your understanding of the subject matter if you follow this advice.

Practice Velvety-Smooth Breathing

Imagery comes more easily when you are in a relaxed but alert state of mind. One method for attaining such a state is Velvety-Smooth Breathing. After reading these next few lines of instruction, close your eyes and keep them closed for the next 10 minutes. Don't look for any images. You'll just aggravate yourself if you can't find them. Focus instead on your breathing. Breathe in and out so smoothly that there is no pause between the in breath and the out breath. Your breathing should be just one, long, continuous, flowing b-r-e-a-t-h-e, like a slow, sensuous sigh. Let it stroke you as you might stroke a smooth piece of velvet.

Start with a Familiar Image

In your first practice run, try describing a familiar person or object in great detail. With your eyes closed, describe your mother, your child, or your spouse. Describe the Taj Mahal or some other grand and famous building.

If you succeed in performing this simple task, congratulations! You have just started working with mental imagery. Many people will deny that they saw an image, but it is psychologically impossible to describe a person or object from memory without first forming a mental image of it.

Practice this skill until you feel comfortable with it. Now you are ready to experience *spontaneous* imagery.

Stay Alert

When you are awaiting spontaneous images, you must be ready for anything. Don't be like Bob S., who suppressed the image of the automobile tire because he thought he was supposed to see something else. You're not "supposed" to see anything. Imagery can come in any form. It might be a fence, a face, or a tree branch. It might be the feel of touching sand, a whiff of gingerbread, or even an emotion. It might be as slight as a splotch of color, a few crisscrossing lines, or a pinpoint of light.

The important thing is to stay alert. The moment an image or impression congeals in your awareness, describe the dickens out of it! Many people fail at this point because they think the image must remain in their conscious view the whole time they're describing it. Not so. Even if the image flickers for a second and disappears, you can still keep describing it from memory, just as you described the Taj Mahal. Indeed, the very act of describing it will usually bring it back into view.

Don't worry about accuracy, either. It doesn't matter if you fudge a little bit. Feel free to enhance, exaggerate, or make up parts of your description, as long as your embellishments give the image more vividness and life. Remember to fudge in all five senses. Sometimes noting a certain smell will provoke a visual image, or a sound will remind you of a taste. Especially when you're just starting out, such conscious meddling with the Image Stream may help to kick start a truly spontaneous thought flow.

At this point, put down the book, close your eyes, turn on your tape recorder, and try to Image Stream, relaxing first with Velvety-Smooth Breathing.

Insomniac's Special

Remember that the first commandment of Image Streaming is to describe the images *out loud*. Many beginners think they know better. When they bother to describe the images at all, they do so silently to themselves. This is one of the surest ways I know of to fall asleep. In fact, if you are troubled by insomnia, I strongly recommend silent Image Streaming as you lie in bed. It will ensure you some snooze time better than anything in your medicine chest.

When he was engaged in his "deep thought" experiments, Einstein would hold a rock in either hand. If he started to drift off, the rock would fall and snap him awake. This method will certainly keep you from falling asleep, but it will also deprive you of the extra dimensions of speech and auditory feedback essential to Image Streaming.

Some methods for making your Image Stream more vivid and useful follow.

USE THE PRESENT TENSE

Remember that the Third Commandment of Image Streaming is to use the present tense. Even if the image has already vanished, you should never say "I *saw* such-and-such" in your description. Always phrase it "I *see* such-and-such" or "I am looking now at such-and-such."

The Feedback Loop

The Image Stream is self-reinforcing. Almost any stimulus will serve to trigger the stream of images, but from that point on your own flow of verbal description is what keeps the Image Stream going. In general, *the more you describe something, the more of it you get*. Your descriptive monologue forms part of a feedback loop. It both *follows* your Image Stream and helps *create* it (see Figure 3.1).

This is why fudging or making up parts of your description is okay. The fudging is a legitimate imaginative process that feeds back to the unconscious and blends with your more spontaneous imagery. Much like the deliberations of an artist or novelist, your fudging is guided largely by the unconscious, even though it appears to be subject to your will. Likewise, your use of the present tense feeds back to the Image Stream, making it more immediate and durable.

Try another Image Stream now, this time staying in the present tense. See what effect this has on the immediacy, power, and clarity of your experience.

Figure 3.1 Describe a perception or observation aloud to an external focus, such as a tape recorder or another person, and examine your perceptions as you speak. This process draws ever-larger portions of unconscious thought into focused attention, which then feed back into your consciousness.

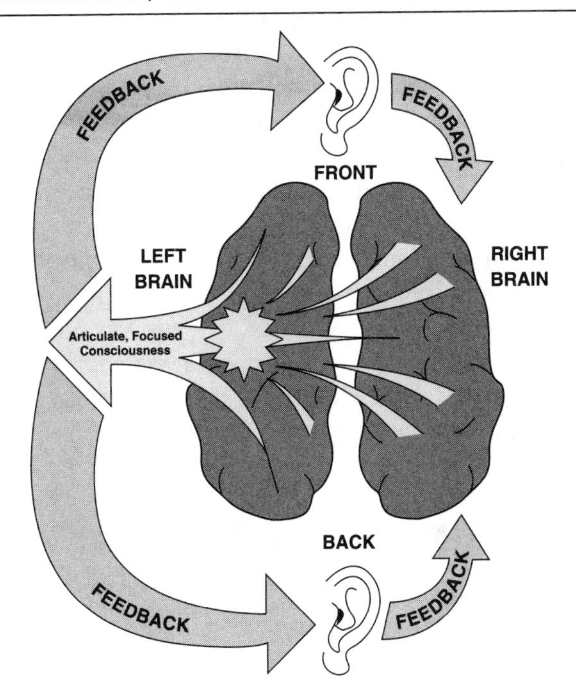

IT'S A FEELING

Beginning in the 1960s, scientists used biofeedback to train people to exercise yogilike control over their bodies. Test subjects spent hours viewing readouts of their heart rate or skin temperature while trying to change the data through force of will. To the astonishment of the experimenters, subjects quickly learned to alter heart rate, skin temperature, and other supposedly involuntary functions.

The same principle is now being applied to create "brain-actuated" machines. Experimental subjects have succeeded in playing video games and working computer programs using only their brain waves, as transmitted through electrodes taped to the scalp. These test subjects cannot explain how they learned to manipulate their brain waves so subtly, nor can they teach anyone else to do it. "It's just a feeling," they'll tell you. You can develop this feeling only by reacting to the feedback. When you think one way, the computer or video game responds. When you think another way, you get no response. Through trial and error, you eventually learn what works.

The skills of Image Streaming are acquired in the same way. As you practice, you'll find that certain images are naturally clearer and longer-lasting than others. In what is perhaps more than coincidence, such images also happen to be the ones that are most beautiful and enticing. Let instinct be your guide. If you are drawn to an image, go with the flow. Hold onto that image as long as you can. The attractiveness of the image interacts with your attentiveness in yet another powerful feedback loop, reinforcing your ability to generate ever-stronger mental imagery.

THE PRINCIPLE OF DESCRIPTION

The above phenomena can be summed up in a three-part Principle of Description:

1. When you describe any object, real or imaginary, *at the same time you are observing it*, the very act of description focuses your attention in such a way that you perceive more and more detail about the object being described.
2. Describing an object aloud to an external focus, such as a live listener or a blank tape in a cassette recorder, is the strongest way we know of so far to build this "additional discovery" effect.
3. The more sensory your description and the less abstract and explanatory, the more powerful the effects, especially when you are describing abstract or complex situations (as distinct from concrete ones).

We return later to this Principle of Description. For now, suffice it to say that this principle lies at the root not only of Image Streaming but of genius itself.

STREAM OF CONSCIOUSNESS

As you work on your Image Stream, never forget that your ultimate goal is to achieve a completely spontaneous flow of uncensored, uninhibited imagery not unlike the free association that psychoanalysts encourage in their patients. At a certain point in your practice, you will no longer have to struggle and strain to get an image. Instead, you will be surprised by the abundance of images, their startling clarity, and their bizarre and unexpected subject matter. Only when this happens have you truly begun to Image Stream.

A TYPICAL IMAGE STREAM

At my suggestion, coauthor Richard Poe produced the following Image Stream for demonstration purposes. Richard's references to the Aha! effect and to Thresholding relate to problem-solving techniques that will be covered in later chapters. We offer this transcript, uncensored and unedited, to give you an idea of what a raw, descriptive monologue looks like.

> I see a kind of background of greenish yellow fuzzy spots, like the spots of a jaguar, or a cheetah, all over a black background, shimmering, and I move closer to them and they get bigger. I'm zooming in on them, the spots are getting bigger, the black spaces are in between, and I see that indeed it is like the fur of a jaguar. I can reach out my hand and I can touch it and I can feel the smoothness against my hand and I can see the cheetah's head. It's really a cheetah, not a jaguar. They're the spots of a cheetah. It turns to me and I can touch its head. I can feel its ears, and their floppiness, the rubbery floppiness and feel the drool of its mouth as I stroke it all over its head and it looks at me, regards me warily, not warily but in an accepting way, as if I'm part of its cheetah family, and it looks out over the savannah, perhaps looking for game, and it's, we're on the African savannah. The trees are very sparse. I can look out and there's a waterhole out there with zebra and wildebeest and a bright sunlit scene, a grassland that stretches out over the horizon. The wind is very hot against my skin, and I feel at one with the cheetah. We're together and we belong together, and I'm not a stranger in this place, and we're hunters together surveying the herds that are out there, and I'm just feeling the wind. It's very hot. It's high noon, I think, in tropical Africa, and the sun is very severe. It's burning in the sky. I can't look at it because it's too bright, but it's almost directly above us, and it's just burning and blazing. It's so bright that the sky doesn't really look blue. It's kind of a sickish yellow, and . . . I have to stop to make neurological contact with this scene again . . . the wind and the sun and the great expanse, looking out, I don't find an "aha," I don't find an answer to my question. My question, of course, is how to, how best to approach the writing of the book. As I think of that, the sky seems to open. A great figure, a man I think, I can't tell if it's a man or a woman, it's a woman with dark hair and a bodice, her hair tied up in a fashion, and dressed in a fashion that evokes for

me Renaissance Italy. Her hair is dark. She looks Italian. She's middle aged, I think. I think of one of the Borgias perhaps or one of the great Renaissance families of Italy, that she's one of them, royalty perhaps, a countess. And she looks at me with very piercing eyes. I can smell the perfume that wafts from her, the scent in her hair. I don't know what to call it but it's a basic scent like frankincense, something old and unsophisticated, something that would have existed in that day. Behind her is a room. She's still standing with her hands pushing back the torn edges of the sky from which she emerged, and behind her is a room and it's musty in the room. I'm going past, I'm going into the room, and the air is musty and close, very warm. There are beakers and flasks and bottles, and there is an alchemist's kiln—an athanor—made of brick. And there's a fire. It's glowing very hot, and the hot wind blasts me, and I'm sweating because it's so hot. It was hot in the African savannah and it's hot in this room, very close and hot and I look down into the fire in the athanor. I look deep into it and I see a baby, almost like an embryo curled up within the fire, and I know it's giving birth to something, of course, but what's in the fire? I look at the baby. It looks more and more like an embryo as I look. The sweat is pouring down my face and I feel that it's purging me and that the poisons are leaving my body as they pour out of me, and I can smell the burning fire. I can smell the fire as it heats the brick, a pungent smell. And I can hear it roaring within. I can hear the power of the furious heat inside of it. As I look deep into the hot part where it's burning, it's all glowing orange and bright, and purplish flashes dart through it and I see something like a ball, a purplish ball. It's like a crystal ball, and I've decided that I'm going to do a thresholding technique, for when that ball bursts open, the answer will lie within it. It's growing black and gnarly and it's seething like an egg, like a reptilian egg with something in it. It bursts and, of course, a dragon, a black dragon in typical medieval form with wings, and it stretches its wings and it flies, and it flies out from the earth off into space and it flies out into the galaxy, into the solar system, between the planets and I'm riding on its back out into deep space. . . .

WHAT IF IT STILL DOESN'T WORK?

By this point, most readers have put down the book at least once and attempted to Image Stream. At least 30 percent have had dismal results. Some may have become so discouraged that they've already decided, "I'm just one of those people who can't get images."

In fact, no such people exist. *Everyone* can visualize. Until 1973, I myself was one of those people who stubbornly maintained that they were not visualizers. Only through an intensive regimen of training and experimentation, which included formal instruction in autohypnosis, was I at last able to get pictures. Once the Image Stream started flowing, however, I very quickly found myself to be "fluent."

No one reading this book will have to go through what I did. The fruits of my twenty years of research and struggle are fully contained in these pages. In the remainder of this chapter are listed some of the most powerful techniques that I have discovered. They will help you break through blocks and start your Image Streams. Since these techniques were compiled and perfected—a process I completed in about 1982—not one out of all the thousands of people who have gone through my seminars has ever failed to get pictures.

The Beautiful-Scene-Describe-Aloud Technique

Beauty is the native language of the right brain. Nothing will reinforce visual thinking more directly or powerfully than the apprehension of beauty and wonder.

Begin your Image Streaming session by recalling the most beautiful natural landscape you have ever seen. It might be a forest, a babbling brook, a garden, or a sunset over the ocean. But it should be a real place, not an imaginary landscape. Then you can focus your attention on recollection rather than fabrication, and you can draw on latent memories about that scene that you have stored and probably forgotten for years.

As you go deeper into the description, your mind will wander and start generating unrelated images. Go with the flow—pick up on these new images and describe *them*.

The Afterimage Technique

Stare at a fairly strong light source, such as a 40–60-watt bulb—but definitely not the sun!—for a half minute or so. Then close your eyes and describe the afterimage of that light source. The afterimage will soon begin to change colors, shape, and position. New images will appear. Keep describing whatever you see, and you will soon drift into a full-fledged Image Stream. Strong visual patterns, such as stripes, polka dots, and checkerboards, will also create afterimage effects if you stare at them long enough.

A variant of this method is the Phosphene Technique. Gently rub your eyes like a sleepy child and describe the lights and colors that result from the pressure of your fingers.

The Old Dream Recall Method

All of us have had unusually vivid dreams that stick in our minds. Some of them we remember all our lives. Try Image Streaming from one such dream. Describe it as vividly as possible, in its original narrative sequence. As with ordinary Image Streaming, don't worry if you have to fill in gaps by fudging a little bit and making things up. If you keep describing everything that happens in the present tense, as if you were looking at it now instead of recalling it, you will soon find yourself looking at real-time images of what you are describing.

At a certain point, the dream will take on a life of its own, resuming where your memory left off and completing itself in your consciousness. You may even rediscover the dream's original message or ending, which you may have forgotten or repressed over the years.

The Story Method

You can build an Image Stream from any good, entertaining novel, story, film, or TV program. It should be a story that you have read or encountered very recently or one that has remained vivid in your mind. Begin "word painting" scenes from that story that were not described by the author. As your description picks up speed and intensity, you will start free-associating into a more genuine independent Image Stream.

The Fantasia Method

This technique stimulates the synesthetic response more directly than any other technique. Listen to some richly textured music with your eyes closed. Experience has shown that French Impressionistic music, classical music (strictly speaking, from 1750 to 1825), and progressive jazz are among the most effective. Density and complexity are key factors. You want to make sure there's enough "music per unit of music" to activate your more sensitive faculties. Such music will stimulate visual images as pleasing and fanciful as those that appear in Disney's *Fantasia*.

The Blindfolded Grope

More tactile-oriented readers may achieve better results from the Blindfolded Grope. One variation is to blindfold yourself and walk around your house feeling different objects. Describe at length the appearance of each object you feel. Another variation is to have someone put together a mystery grab bag of diverse objects for you. Remove and feel each object in turn, describing as you go.

Eating or Smelling Blindfolded

This method helps involve the much-neglected faculties of taste and smell. Blindfold yourself before eating. While you eat, describe in detail every sensation, until an unrelated Image Stream begins.

Another variant is to select four or five spices from your kitchen with a range of pleasing aromas. Pull out the stoppers and set the spices in a row before you. With your eyes closed, shuffle them around and try to identify each one by smell. You may be surprised at the intensity and diversity of the Image Streams the odors conjure up. Smells are particularly effective at evoking long-forgotten memories.

Air Sculpting

Close your eyes and start sculpting with your hands some objet d'art out of thin air. When it's finished, hold the imaginary sculpture in your hands and describe it in detail. Making a real sculpture out of clay will also work, if you don't mind the mess. Like the Blindfolded Grope, this technique may work better for more tactile readers.

The Commuter Special

While riding in a train, bus, or car, close your eyes and start describing from your imagination the landscapes and street scenes you think you are passing. Don't be embarrassed to describe the scenes aloud. When other passengers see your tape recorder, they will assume you are a high-powered executive dictating letters. This is a great way for busy people to convert downtime into a mind-expanding experience.

Tree and Cloud

Imagine that you are walking in a meadow. You find yourself going up a hill with a single immense tree at the very top. Engage all your senses. Exult in the warm breeze; the sunshine on your face, neck, and shoulders; the smells of the meadow; the muscular pull of walking up a gradual slope for a long time. Consider the variety of wildflowers, the swish of your feet through the grass, and the sound of your own breathing.

When you reach the top of the hill, lie down in the soft, cool moss at the base of the tree. Look up the tree's immense trunk. Look through its branches at the sky, where clouds are scudding across an azure expanse. Notice how the movement of the clouds creates an illusion that the tree itself is moving, along with you and the hill. Now drift with that movement. Let it take you wherever it will.

Beneath the Boat

Imagine riding a boat onto a gentle lake. Peer down into the water, past the sparkle and the ripples. Try to make out what's down there. At first, you see only the play of refracted sunlight across the bottom, but as you peer more intently, a whole underwater world takes form. Your own imagination will determine whether you see submerged Atlantean civilizations or fabulous sea creatures. Describe whatever you see in rich detail.

The Time/Space Method

This technique led Einstein to his Special Theory of Relativity. Recall from Chapter 1 that Einstein spent ten years of his life trying to imagine what it was like to move at the speed of light. You can try something similar.

Imagine yourself to be some sort of electromagnetic or gravitational phenomenon, such as a radio wave, a laser beam, a quantum particle, or even a black hole. Imagine that you are moving across deep space, between stars, between galaxies, into the unknown. You don't have to be a physicist to achieve profound insights through this method. But boning up on a few popular science books will fuel your imagination and enrich the experience.

Live-Partner Feedback

If the above techniques don't work, you may need to enlist the help of a live partner. This method, sometimes called the Helper technique, is based on the assumption that you really are Image Streaming all the time but simply aren't consciously aware of it. Your partner's job is to help you become aware.

Use the normal procedure to start Image Streaming, including the Velvety-Smooth Breathing technique. Instruct your partner to watch you closely as you Image Stream, following these instructions. In the following four steps only, the word "you" refers to the partner, not the Image Streamer.

1. ***Look for attention cues:*** Attention cues are tiny indications that something has caught the Image Streamer's attention. A momentary pause in breathing is a dead giveaway, especially when the person is practicing Velvety-Smooth Breathing. It means he has reacted to some inner stimulus, such as a striking mental image. Another good clue is eye movement under the lids. When the Image Streamer's eyes are closed, what could he possibly be tracking? It can only be mental imagery. Remember that you are looking for real eye movement, not a mere flutter of the eyelids.
2. ***Alert the Image Streamer:*** The instant you spot an attention cue, alert the Image Streamer by asking, "What was in your awareness just then?" When in doubt as to whether you really saw such an attention cue, go ahead and ask anyway—it can't hurt.
3. ***Persist:*** Some Imagers may have such powerful blocks that they refuse to believe they are really Image Streaming, even after you've cued them a dozen times. Be patient. Persist in the exercise until it takes.
4. ***Coach the Image Streamer:*** At some point, the Image Streamer will finally catch on. He will announce, with great excitement, that he has just gotten an image. At that point, you take on a new role as your partner's tape recorder or coach. It's your job now to coach your partner to continue describing each image, in present tense and

in rich sensory detail. Keep reassuring the Image Streamer that it's okay to keep describing for several minutes even if the image flashed for only a second. To encourage the flow, you can remind him that it's okay to fudge a little and to make up details. In the flush of excitement from his first real image, the Image Streamer may well have forgotten these liberating principles and will benefit from timely reminders.

Co-Tripping with a Live Partner

If you are working with a live partner, you may wish to try co-tripping. This is a slightly more advanced technique, and you should undertake it only after both you and your partner have become adept at Image Streaming.

Sit down facing each other, and both of you close your eyes. Each of you should then start describing to the other your respective images. Instead of waiting to take formal turns, when one of you pauses for breath, the other rushes in. Allow no empty air time to elapse once you are under way. The gamelike quality of this exercise is an excellent catalyst for Image Streaming. It will rapidly sharpen your imagery and descriptive powers.

STRENGTHEN YOUR CONTACT WITH THE IMAGE STREAM

Once you've opened contact with your Image Stream, you will want to test and explore its limits. The farther you push your powers of perception, the stronger your imaging faculties will become. Below are a few suggested exercises. These and many other similar procedures not only build up your imaging muscles, but actually increase the apparent size of the imaginary space in which Image Streaming takes place. By carving out a larger, panoramic stage, you free up your mind to create ever more elaborate and exhilarating dramas.

The Panoramic Scan Technique

As soon as a clear scene comes into view, begin panning slowly to the left, as if your head were a camera and your verbal description the videotape. As you turn your head, what new details of the scene come into view? Turn completely around. What do you see behind you? When you have completed a 360° pan of the entire scene, you will feel an extraordinarily realistic sense of standing in three-dimensional space.

The Expanding Senses Technique

Vision is the strongest of all the senses, and you will tend to rely on it disproportionately. Practice describing your Image Stream using your nonvisual senses exclusively. Touch is probably the most important nonvisual sense. Explore the different surfaces in your imagined scene. Feel the bark on the trunks of trees, the roughness of a brick wall in the sun (which, by the way, also has an unusual characteristic smell), the dew-fresh grass underfoot, the grain and sheen of fine wood furniture, the texture of plush carpeting. Also valuable for this purpose are warmth and coolness, moistness, atmospheric feel, sense of space and motion, and mass and lightness.

Freeing Up Your Point of View

You can change your spatial point of view by moving around in the imaged scene, describing whatever comes into view as you move. You can also go back and forth in time by changing the time of day, the time of year, the century, or the millennium. Finally, you can change your size, growing larger and smaller like Alice in Wonderland. Explore to discover additional ways to perceive even more detail in your Image Stream.

THE DA VINCI PRINCIPLE

As you practice these techniques, you will no doubt discover many tricks of your own for opening the Image Stream. Your methods may even work better than those I have presented. There are no hard and fast rules. Indeed, a cardinal principle of Image Streaming is that *it doesn't matter how you get the images started.* Once the stream starts flowing, it takes you where it will. It transports you swiftly to the precise psychological space in which you most need to be.

I call this the da Vinci principle. Psychologists today know that virtually any stimulus—such as the meaningless ink blots of the Rorschach test—can set off a train of association that will lead rapidly to the most sensitive places in your mind. Leonardo da Vinci discovered this principle 500 years before Sigmund Freud. Unlike Freud, da Vinci didn't use free association to seek out his deep-rooted complexes. Instead, the great Renaissance Man of fifteenth-century Florence free-associated his way to artistic and scientific insights.

"It should not be hard," wrote da Vinci in his *Notebooks*, "for you to stop sometimes and look into the stains of walls, or ashes of a fire, or clouds, or mud or like places, in which . . . you may find really marvelous ideas."[8]

In a treatise on painting, da Vinci suggested that such random forms, when studied closely, soon take on a "resemblance to various different landscapes adorned with mountains, rivers, rocks, trees, plains, wide valleys, and various groups of hills. You will also be able to see divers combats and figures in quick movement, and strange expressions of faces, and outlandish costumes, and an infinite number of things which you can then reduce into separate and well-conceived forms."[9]

Da Vinci also found inspiration in the sound of bells, "in whose clanging you may discover every name and word that you can imagine."

You may feel a little silly practicing some of the techniques in this chapter, but don't worry. You're in good company. Da Vinci, too, acknowledged that cynics might find humor in his "new device."

"It may appear trivial and almost ludicrous," he wrote. But the method "is nevertheless of great utility in arousing the mind to various inventions."[10]

THE WHITMAN-BLAKE EFFECT

Two hundred years ago, the poet William Blake penned four lines that, to me, embody the essence of Image Streaming:

To see a World in a Grain of Sand
And a Heaven in a Wild Flower
Hold Infinity in the palm of your hand
And Eternity in an hour[11]

Walt Whitman echoed the same idea a hundred years later when he wrote that one may "see the universe in a single blade of grass." The tiniest stimulus will indeed provoke thought-storms and reveries without limit. Through a process I call the Whitman-Blake Effect, your limbic brain works its synesthetic alchemy upon the raw input of your senses, transmuting it into genius.

Pause before making your next big decision. Look around and notice the slight irregularities of the ceiling, the texture of brick underfoot, the feel of your knee bending and straightening, and the slight shifts of sensation in your shoulders, stomach, neck, and face. You can't really explain why, but when you widen your neurological contact with the world in this way you feel stronger, wiser, and more creative—and you choose more wisely.

Image Streaming amplifies this effect a hundredfold. When you master the techniques in this chapter, you will have acquired a powerful microscope through which you can indeed spy worlds and universes in the tiniest sand grains and grass shoots of your mind.

CHAPTER 4

AMPLIFY YOUR FEEDBACK

As a student in Zurich during the late 1890s, young Albert Einstein often partook of sailing excursions on the Zurichsee. One of his early companions, a young woman named Fräulein Markwalder, later recalled Einstein's peculiarly unsociable behavior during these outings.

She noted that every time the wind died and the boat lay becalmed in the water, Einstein would immediately pull out a notebook and start writing, oblivious to his companions.

"But as soon as there was a breath of wind," Fräulein Markwalder remarked, the notebook vanished and young Einstein "was immediately ready to start sailing again."

What was Einstein writing in his notebook? Perhaps we will never know. But far more important than *what* he wrote is the question of *why* Einstein wrote. It can hardly be an accident that researchers in the field of high intelligence have long regarded the habit of compulsive scribbling as one of the telltale hallmarks of genius.

COMPULSIVE SCRIBBLERS

In the 1920s, researcher Catherine Cox studied 300 geniuses from history, such as Sir Isaac Newton, Thomas Jefferson, and Johann Sebastian Bach. Her exhaustive survey of available biographies revealed a pattern of strikingly similar habits and personality traits among these top achievers.

One sign of genius, Cox noted, was a penchant for eloquently recording thoughts and feelings in diaries, poems, and letters to friends and family, starting from an early age. Cox observed this tendency not only in budding writers, but in generals, statesmen, and scientists.[1]

The 1 Percenters

A casual trip to the library will confirm Cox's finding. It has been estimated that fewer than 1 percent of the population habitually engage in writing out their thoughts, experiences, and perceptions, whether in journals, diaries, letters, or books. But, with startling consistency, the world's top achievers seem to fall in that critical 1 percent.

At the library, you will notice the vastly disproportionate amount of autobiographical material written and published by history's most gifted men and women down through the centuries, among them Benjamin Franklin's famous *Autobiography*, Einstein's *Autobiographical Notes*, and Leonardo da Vinci's voluminous *Notebooks,* filled with sketches, diagrams, and cryptic writing. Thomas Edison produced some 3 million pages of notes and letters before he died in 1931.[2] The question is, does genius lead to scribbling, or does scribbling lead to genius?

Why did these gifted men and women start keeping diaries in the first place? Was it because they knew in advance that they would someday be famous and wanted to leave behind a record for future historians? Was their writing simply an irrelevant by-product of a highly expressive mind—or a highly inflated ego? Or—and this is the point I will argue here—was the scribbling, in and of itself, a mechanism by which people who were not born geniuses unconsciously nurtured and activated a superior intellect?

AN ANCIENT SECRET

In a famous scene from the 1959 film classic *The Nun's Story*, Audrey Hepburn and the other postulants line up to receive small, leather-bound notebooks.

"For the rest of your lives," says the Mother Superior, "you will examine your consciences twice a day and write your reflections in these notebooks."

For Audrey Hepburn, the diary proves a burden. Her mind wanders as she writes. Perhaps she would have been more diligent had she been privy to recent neurophysiological discoveries that suggest this ancient spiritual practice may harbor potent secrets for enhancing brainpower.

Super Nuns

For years, gerontologist David Snowdon of the Sanders-Brown Center of Aging at the University of Kentucky has been studying an obscure community of nuns living in Mankato, Minnesota. Like others of their order—the School Sisters of Notre Dame—the sisters of Mankato live an unusually long life. Twenty-five of the 150 retired nuns in the Mankato convent are over ninety, and a few have passed the century mark. Of even more significance, they are remarkably resistant to the usual brain diseases afflicting the elderly, such as Alzheimer's, stroke, and dementia, which strike the sisters of Mankato far less often, less early, and less severely than they do others.

What is the secret of the Super Nuns? Snowdon intends to answer this question. Three years ago, 678 Mankato sisters agreed to donate their brains for Snowdon's study. So far he has collected ninety-five brains. When his study is complete, Snowdon predicts that a large portion of Mankato nuns will show an unusually rich growth of interconnections between neurons in their brains. Old age and diseases like Alzheimer's tend to block and shrivel these pathways, but if you have more than enough to spare, your brain can use the extra dendrites and axons to bypass damaged areas.

Use It or Lose It

We have already learned that dendrites, axons, and glial cells multiply in response to mental challenges. It is, in fact, to the rigorous intellectual regimen of the Mankato nuns that Snowdon attributes their robust neurophysiology.

Even more than others, their religious order condemns the sin of mental idleness. They forbid their brains the luxury of downtime. Many of the sisters pursue higher degrees. They play quiz games and solve brainteasers to pass the time, and they debate politics in weekly seminars. Most important, each nun keeps a detailed journal of her personal spiritual quest. Like Audrey Hepburn in *The Nun's Story*, they examine their souls daily and record on paper what lies within.[3]

THE HANDS-ON PRINCIPLE

When Marian Diamond carried out her famous experiment of putting rats in a super-stimulating environment, she also tested a control group. This control group was not allowed to play with the toys, swings, ladders, treadmills, trapezes, and other delights enjoyed by the Super Rats. They were, however, allowed to *watch* the Super Rats play.[4,5]

Certain theories of child development hold that stimulating input alone will enhance a child's intellectual growth. By analogy, Diamond's spectator rats should have sprouted extra neurological connections just by watching the other rats play—but they didn't. The spectator rats died just as young and had interconnections just as sparse as did their less fortunate counterparts in barren cages who were not allowed to watch.

Clearly, input alone did not create Super Rats. The rats had to touch and play with the toys in order to gain brainpower. This suggests that the Super Rat Effect was a feedback loop. The more the rats physically interacted with their environment, the more stimulation that environment fed back to them in the form of brain growth.

THE EXPRESSION CIRCUIT

Diamond's finding dramatically underscores something neurologists have known for eighty years: A sizable portion of our brains' physical development depends not on genetic inheritance, or even on outside stimulus, but rather on the *feedback from our own spontaneous and expressive activity*. I call this the Expression Circuit.

Santiago Ramon y Cajal noted in 1911 that the microneurons of the cerebellum develop in direct response to an infant's activities.[6] More recently, neurophysiologist José M. R. Delgado has observed that up to 90 percent of the neurons in certain parts of the brain take form after birth, their number and structure heavily influenced by "sensory inputs from the environment."[7]

Although this powerful feedback loop has been recognized for over eighty years, its implications for accelerated learning have long been overlooked. By harnessing the power of self-expression and sensory feedback, we can actually change the physical form of our brains.

Creeping and Crawling

A psychologist once studied two American Indian tribes who lived so closely together that their reservations were physically intertwined. The tribes shared similar cultures and economic circumstances. Yet one tribe tested with an average IQ over twenty-five points higher than that of the other. This tribe, the researcher noticed, allowed its infants to creep and crawl freely, while the tribe with lower average IQ restrained them.[8]

Studies by Dr. Raymond C. Dart's Institute of Man in Philadelphia have included similar observations. Dart discovered *Australopithecus*, an important missing link in human evolution. His institute now studies general human development. It has found, from intensive study of cultures at all levels of complexity, that those tribes whose infants creep and crawl tend to have more complex societies, higher technology, and some form of written language. Most tribes that restrict their infants from crawling have no writing of their own and can be taught to read only with great difficulty. Moreover, people raised in such tribes have remarkable difficulty focusing their eyes at arm's length—the very distance at which people read, write, perform arts and crafts, and make and use tools. Truly, "civilization was built at arm's length," as accelerative learning pioneer Dr. Glenn Doman once remarked.

The Hand-Eye Circuit

Watch a baby's eyes as it creeps about the floor on its hands and knees. As each hand moves forward in turn, the baby's eyes focus on it, back and forth, following each hand, many thousands of times. The baby is not only learning to crawl but also training its eyes to work together at arm's length. Each time the baby *expresses* its will by reaching out a hand, it receives *feedback*—through the touch of the rug, the sight of its hand executing the brain's orders, and the sense of achievement that no doubt comes from successful locomotion across the floor.

The Expression Circuit model, when it is more widely understood, will do much to quell ongoing arguments about what constitutes a stimulating environment for a growing child. A conventional method of enriching a baby's environment might include stringing flashing, multicolored Christmas lights around the newborn's crib. These lights will certainly stimulate the baby's vision, but they provide little or no feedback. They will not help the infant's brain growth much more than watching the Super Rats play helped the spectator rats.

A model more in keeping with the Expression Circuit model would allow the infant to control the Christmas lights, say, from pressure pads in the crib. The infant would thus gain instant visual feedback from his or her own movements. Brightly colored mobiles that move and shimmer when they are swatted are also a good way to provide feedback to older infants.

HOW STEPHEN HAWKING CHEATED DEATH FOR 30 YEARS

The Feedback Principle may have a lot to do with why renowned physicist Stephen Hawking is still alive and working. After diagnosing Hawking with Lou Gehrig's disease, a degenerative neurological illness, his doctor gave him two years to live. Today, thirty-two years after that grim prognosis, Hawking not only lives but has recently remarried and continues to set the pace in the field of global general relativity theory, the most advanced realm of modern physics. He even became an "actor" by playing a role in person in an episode of *Star Trek: The Next Generation*!

How does he do it?

Confined to a wheelchair, unable to write, Hawking for many years spoke only in mumbles. Now, having lost his voice entirely, he communicates by painstakingly tapping out on a keypad messages that are then "spoken" by an electronic voice. Hawking has thus spent many years surrounded by attentive graduate students who hang on his every word and jot down every thought—a stimulat-

Figure 4.1 Physicist Stephen Hawking was given two years to live by his doctors, due to a degenerative neurological illness. But thirty-two years later, he continues to produce the most brillant cosmological insights of our day. Hawking constantly throws out ideas and receives instant feedback from attentive graduate students who hang on his every word. This intensive feedback loop may have kept him alive for thirty years. Neurologists attribute a large portion of brain development—physical and mental—to such feedback loops.

ing cycle of self-expression and feedback that may have kept him alive and lucid for the last three decades (see Figure 4.1).

THE PHYSICIST WHO COULDN'T UNDERSTAND MATH

The English inventor Michael Faraday was one of history's greatest scientific minds. His theory of electromagnetic fields and lines of force largely inspired Einstein's quest for relativity. Yet Faraday's methods have long puzzled the more strait-laced historians of science.

"Faraday . . . was completely innocent of mathematics," marveled Isaac Asimov in his *History of Physics*; " . . . he developed his notion of lines of force in a remarkably unsophisticated way, picturing them almost like rubber bands."[9]

Scientists would probably never have known what to do with Faraday's electromagnetic fields had not James Clerk Maxwell later set them into math.[10] Poor Faraday did his best to follow Maxwell's work, but he eventually grew so befuddled that he wrote Maxwell, beseeching him to "translate" his "hieroglyphics" into the sort of "common language" that "such as I" could understand.[11]

Faraday's Notebooks

Even less conventional were Faraday's notebooks. He filled thousands of pages of notes and diaries with wild, unstructured—but often quite poetic—streams of thought.

"Today, every fall was foaming from the abundance of water," he wrote in his diary in 1841 after hiking near the waterfalls of Giessbach in Switzerland. "The sun shone brightly, and rainbows seen from various points were very beautiful."[12]

One rainbow in particular caught Faraday's eye.

"It looked a spirit strong in faith and steadfast in the midst of the storm . . . ," he wrote. "And though it might fade and revive, still it held on. . . . And the very drops, which in the whirlwind of their fury seemed as if they would carry all away, were made to revive it and give it greater beauty."

As science writer Thomas West noted, it is perhaps more than coincidence that the phenomenon that so entranced Faraday—the persistence and strength of the rainbow's form in the midst of the waterfall's raging chaos—resembled so closely the mysterious manner by which lines of electromagnetic force preserved their shape in the formless "ether."[13]

One frustrated scientist, attempting to reconstruct Faraday's scientific thought process from his writings, was appalled to find nearly every page filled with such seemingly aimless "Image Streaming."

"The *Diaries* have the . . . irritating form of ideas jotted down, repeated and forgotten," he wrote. Try as he might, this scientist searched in vain for anything that smacked of sustained reasoning. Instead, he encountered a "morass of articulated and unarticulated principles, concepts, observations and physical facts." At length, this hapless scholar was forced to conclude that the very "*lack of pattern* was . . . *itself the evidence of how Faraday thought* [emphasis in original]. . . . Faraday suspended the need to understand," he concluded, "and simply acknowledged the thoughts which came into his head. The coherence of ideas was not imposed by any prior framework, but was allowed to emerge from the chaos of thoughts he experienced."[14]

THE PORTABLE MEMORY BANK TECHNIQUE

Faraday was using a technique I call the Portable Memory Bank. It's quite simple. You buy a small notebook and carry it around with you wherever you go, just as young Einstein and Faraday did. Write down in that notebook any stray thoughts that come into your head, whether or not they seem worth recording at the time.

The technique works because of the First Law of Behavioral Psychology. Whenever you write down a perception or an idea, you reinforce the behavior of being perceptive or creative. Whenever you fail to describe or record such insights, you reinforce the behavior of being unperceptive and uncreative. Simple, isn't it?

You will not have to practice this technique long before you notice a sharp increase in the number and quality of creative thoughts that pop into your head. In fact, compulsive scribbling is a crude form of Image Streaming. Like Image Streaming, it creates a feedback loop between your mind, which generates ideas, and your self-awareness, symbolized by the writing in your notebook.

THE DRASHTA EFFECT

The Hindus have a special word for self-awareness. They call it *drashta*, sometimes translated as the "looker" or "witness." *Drashta* is purely objective. It looks dispassionately on the world and even allows us to separate from our bodies and look back at ourselves. We can think of the notebook used for our Portable Memory Bank as a method for reinforcing our *drashta* perspective. It allows us to scrutinize our own stream of subtle perceptions as if we were an outside observer.

In Image Streaming, the tape recorder represents *drashta*, completing a feedback loop between our inner perceptions and our objective "witness." But Image Streaming is far more direct and powerful than the Portable Memory Bank. When we write in a notebook, we describe our thoughts only after the fact, when the flow of description can no longer feed back and influence our mental imagery. Moreover, journal writing neglects the faculties of speech and hearing, which are crucial to Image Streaming.

Both techniques—Portable Memory Bank and Image Streaming—tend to bridge certain poles and stimulate certain brain faculties that are neglected by the other. They are most effective when they are used in tandem, especially when both are focused upon the same specific observation.

THE HEISENBERG PRINCIPLE

In 1926, the German physicist Werner Heisenberg determined that it was impossible to measure the trajectory of an electron hurtling through space. Light waves were too big to cast a shadow of the tiny particle, and gamma rays—which have a much smaller wavelength—were too powerful. As soon as gamma rays struck the electron, they knocked it off its course. The very act of observing the electron thus altered its behavior and contaminated the experiment.

Extrapolating from this phenomenon, Heisenberg formulated his famous Uncertainty Principle, which states that there are some things, such as the speed and trajectory of an electron, that we can never know for certain because the very act of observing them changes the data.

"What we observe is not nature," said Heisenberg, "but nature exposed to our method of questioning."[15]

Heisenberg's principle has since been applied as a metaphor to almost every area of knowledge, including psychology. Professional therapists use it to explain why one cannot psychoanalyze oneself. The presence of a live analyst sitting in the room is needed to catalyze and focus the patient's train of association. Freud theorized that the analyst took on the role of someone in the patient's past, through the process of transference. Modern relational therapists ascribe the phenomenon to the personal interaction between patient and analyst. But, these are all just fancy ways of saying that free association works better when it is channeled through a feedback loop. An *external focus*—whether a person, an audience, a notebook, or a tape recorder—completes the loop, drawing subtle perceptions from a person that would not emerge on their own.

BAD FEEDBACK

Unfortunately, feedback loops do not always promote brain development. Bad feedback will stunt the brain as powerfully as good feedback helps it. Ingenious expression is all too often punished in our society with mockery, envy, and adult disapproval.

I once received some exceedingly bad feedback from my sixth-grade history teacher back in 1949, in Harrisonburg, Virginia. I had remarked in class that it seemed unfair for some people to hoard all the wealth while others starved. Wouldn't it be better, I suggested, if wealth were divided more evenly? A more able teacher might have used my provocative question as a springboard to launch into basic lessons on economics. But, no doubt partly because the McCarthy era was just then dawning, this teacher reacted with fear and suspicion.

"That would be communism!" he sputtered accusingly. "You wouldn't want *that*, would you?"

At the age of eleven, I hardly knew what I wanted, but I was perceptive enough to know when I was being bullied. At my first opportunity, I repaired to the school library, where I seized a copy of Karl Marx's *Das Capital* from the shelf and plunged defiantly into its dense and inscrutable pages. Thanks mainly to Marx's soporific style, I avoided early conversion to the proletarian cause. In fact, by the time I received my Bachelor of Arts degree in economics, I had concluded that the free-market ideals of Adam Smith and Henry Hazlitt—with some modifications—offered the fairest shake for all.[16]

Despite the happy ending, the bad feedback that teacher provided could have done serious damage to my budding intellect. A more sensitive—or perhaps less rebellious—child might have been intimidated for life from asking penetrating questions. For reasons I discuss below, bad feedback is one of the principle blocks to intelligence. Image Streaming offers the most effective way I know to correct and counteract its effects.

The Peanut Gallery Principle

"Experts" have become the reigning priests of our day. Nonexperts are sharply discouraged from expressing opinions. Yet many of the great discoveries in every field have been made by amateurs. Consider Einstein, the humble patent examiner who presumed to contradict full physics professors, or Heinrich Schliemann, the import-export entrepreneur who unearthed the lost city of Troy after the experts had declared it a fable.

It is unlikely that any learned scientists in 1831 would have attempted Michael Faraday's famous experiment of that year. Only the wildest hunch moved Faraday to test what would happen if he spun a copper disk between the two poles of a horseshoe magnet. Faraday himself was shocked to find that this device actually generated electricity.[17]

"This discovery was an audacious mental creation," Einstein later remarked, "which we owe chiefly to the fact that Faraday never went to school, and therefore preserved the rare gift of thinking freely."[18]

Most of us at one time or another have felt a strong suspicion that the experts were wrong about some fundamental claim in medicine, politics, economics, or a host of other subjects. We have a healthy, natural drive to jeer from the peanut gallery. Indeed, our American republic is based upon this peanut gallery principle. Our Constitution presumes that ordinary citizens can best judge the common good, while such experts as judges, generals, and career politicians must be kept in check through a sharp separation of powers and a free and vigilant press. Yet much of our training, whether in childhood, in college, or on the job, is designed to beat the Peanut Gallery Principle out of us. We are taught to distrust our own opinions and lean on the opinions of experts.

It is no accident that the greatest geniuses share a profound irreverence toward conventional opinion. The courage to be different is a cornerstone of high intellect. That is the real moral behind the fable of the emperor's new clothes. The emperor's subjects were told that they were fools if they could not see the emperor's invisible robe. By pretending to be wise, the townspeople showed themselves to be the biggest fools of all. The only truly wise head among them belonged to the naive little boy who trusted his own perceptions and dared to proclaim that the emperor was naked.

THE PARABLE OF THE STUCK TRUCK

A truck once became jammed under an overpass because it was too high for the clearance. Police and road crews came and tried to pry it out, but nothing seemed to work. Everyone had a different idea of how to free the truck. At first, they tried to empty some of the load, but that only made the truck lighter, causing it to rise and jam even more tightly against the overpass. They applied crowbars and wedges. They revved the engine. In short, they applied all the typical methods people use for unsticking cars, trucks, and other mechanical devices. But everything they tried only made matters worse.

Then a six-year-old boy came along. He glanced at the truck's axles and noted their height above the pavement. Then he offered the solution: Let some of the air out of the truck's tires! The problem was solved.

Forget What You "Know"

The police and road crews couldn't free the truck because they "knew" too much. Everything they knew about freeing jammed vehicles and machine parts somehow involved prying or yanking things loose through physical force. Likewise, most of our problems are only exacerbated by the things we know. Only when we shift our attention away from what we know to what we are actually *perceiving* are most problems resolved.

This might sound easy, but in fact it's incredibly hard. Most people, when they begin Image Streaming, have great trouble confining their descriptions to their here-and-now sensory impressions. Their tendency is to gloss over their raw perceptions and talk abstractly *about* the images they see. They describe what they know about the images, not what they perceive.

In subsequent chapters, we will show how Image Streaming is used to answer questions and solve problems. Invariably, the most effective Image Streams arise when people focus on detailed, sensory description rather than talking abstractly about the imagery.

Strangled Eggs, Please

I once conducted a rather silly experiment that nonetheless yielded much food for thought. It all started one day when I was ordering breakfast in a restaurant. I told the waitress that I wanted "strangled eggs." Without batting an eyelash, the waitress brought me *scrambled* eggs, never noticing that I had asked for something slightly more exotic.

Intrigued, I repeated this test on seventy different occasions over the next six years—specifically, on those occasions when the restaurant was quiet and the server spoke good English. On sixty-two of those occasions, the waiter or waitress failed to notice that I had said "strangled," despite the fact that I spoke in a strong, clear voice.

A friend of mine who worked in a high-security area of the Pentagon tried a similar experiment. To pass through the security checkpoint, he needed an authorized badge with his name and picture. One day, he folded up a $5 bill so that only the picture of Abraham Lincoln appeared. He then pasted Lincoln's portrait on the badge over his own picture. Although my friend wore this badge every day for the next year, not one guard out of the ever-changing sets of guards on duty ever noticed it.

Set Thinking

These crude experiments reflect a principle that has been proved many times in more sophisticated studies. By and large, people see and hear exactly what they *expect* to see and hear, even if it differs from their actual perception. Psychologists call this "set thinking," because the mind perceives what it is set to perceive.

To some extent, set thinking is a necessary shortcut. We don't have time, for example, to pay attention to every syllable of every word we hear. It is convenient and sensible for a busy waitress to assume that her customers will ask for eggs only in certain set ways.

Yet set thinking also gives rise to bigotry. It enables some people to assume that all Blacks are dangerous or all Whites racist, that all men are brutal or all women hysterical, that all Democrats are unprincipled libertines and all Republicans narrow-minded dolts. Set thinking also clouds the perceptions of scientists and statisticians, who often see only what they expect to see in a given set of data.

Image Streaming seeks to wean the mind from its dependence on set thinking. The more we engage in Image Streaming, the more we reinforce the behavior of relying upon our actual perceptions rather than on our prejudice. We learn to be what I call Original Observers, people who perceive rather than assume.

BE AN ORIGINAL OBSERVER

To be an Original Observer is not far from being a genius. If we encouraged children (and adults, too, for that matter) to ask questions and seek answers freely, their natural genius would astonish us. Indeed, all children are Original Observers up to the age of four, when they start learning to set their minds at a lower level of curiosity.

This phenomenon partially accounts for the notorious instability of IQ scores in children below the age of four or five. At that age, IQ tests are considered unreliable, because a child of such tender years may test at a substandard level of 70 one week and come back the next week with a genius score in excess of 130. Educators prefer to wait until the child has ossified at a particular level before they begin testing. Unfortunately, most children have already learned by that time that it is bad to be an Original Observer, and their IQ scores reflect this idea.

OVERCOMING BAD FEEDBACK

Bad feedback will be with us for a long time, maybe forever. If we can't abolish it, how can we circumvent it? On an individual level, some people manage to turn lemons into lemonade. They use bad feedback as a spur to become even better Original Observers.

When my sixth-grade history teacher scolded me for my question, he was unconsciously trying to train me in set thinking. He implied that questions about social justice were all lumped in a set or category called "communist" and should therefore not be raised. Because I was stubborn, I resisted this training and resolved to seek my own answers.

As it turned out, my childish concern about inequality ultimately shaped my career. It drove me to study economics and to develop a profound appreciation for the liberties assured by a truly free market (something America has yet to institute). It later drove me to realize that self-improvement, through education and accelerative learning, was the truest path to dignity and equal opportunity for all. That same childish concern prompted me to complete a Ph.D. in education, to write my dissertation on the subject of enhancing human intelligence, and finally to spend twenty-five years developing the technique of Image Streaming.

In the end, I must thank that sixth-grade teacher for making me angry enough to go out and seek my own answers. However, most people are not so fortunate as I.

The bad feedback they receive may be far worse, or they may have been less well equipped to fight back. Indeed, the bad feedback they experience may have occurred so long ago or may have been so unspeakably unpleasant that they no longer remember when or how they ceased to be Original Observers.

Is there a way to turn lemons into lemonade, even years after the fact? Is there a way to go back and fix things after they've gone wrong? Indeed there is. My study of the work of Swiss biologist Jean Piaget led me unexpectedly to the key. Piaget's schema of mental development provides a roadmap whereby we can apply Image Streaming to repair past damage with almost surgical accuracy.

JEAN PIAGET'S SCHEMA

Piaget studied his own children to determine a child's normal path of mental development. He noted that children's powers of thought and perception advance through distinct stages: the sensory-motor stage (birth to eighteen months), the preoperational stage (up to seven years old), the concrete operational stage (seven to eleven), and the stage of formal operations (eleven to fifteen). The actual ages vary from child to child.

Children advance through these stages by engaging in prolonged feedback loops with the world around them, learning by trial and error. In the early preoperational stage, for example, at around three or four years old, Piaget discovered that if you pour water from a short, thick glass into a tall, thin one, a child will believe that there is more water in the tall glass because the water level is higher. The child will even be afraid that if you pour the water back into the squat container, it will overflow. He has not yet learned the principle of conservation. But after a certain number of years experimenting with liquids, the feedback loop will teach him the important principle of the conservation of quantity.

Missing Steps

Problems arise, according to Piaget, when a child skips over stages in the normal development, adopting concepts abstractly instead of grasping them intuitively.

If, for example, adults and schools come along and teach the child that the amount of water in the glass remains constant regardless of the container's shape, the child will have a less firm grasp of this concept than if he learned it through experience. He will have skipped an important step in development, and all subsequent stages will therefore rest upon a shakier foundation. That child's perception of himself, of others, and of the world and universe around him, are from that point on, damaged and weakened. His development in later years might be handicapped or distorted.

Obviously, it is impossible—and even physically dangerous—for a child to learn everything by trial and error. Our ability to learn from others' experience is one of the things that makes us human. But such learning must allow the child to rediscover principles through a creative process similar to that of the principle's original discoverer. For example, Max Wertheimer, the founder of Gestalt psychology, pointed out that when children grasp the underlying *reason* why the area of a parallelogram equals its base times its height, they not only remember the formula better than children who learn it by rote but they easily go on to figure out on their own how to calculate the areas of many other geometric shapes.[19]

For this reason, Piaget has been greatly alarmed by the way his theories have been used to justify supercharged learning programs that bombard children with information out of sequence instead of allowing them to learn concepts from experience, in their natural order.

Hypnotic Regression

In my doctoral dissertation,[20] I speculated that people might be able to go back and repair such weak points in their learning sequence through hypnotic regression. A trained hypnotist could regress a person back to a critical moment in his intellectual development and provide some enriching experience to build the key concept that had been taught out of sequence.

Because such hypnosis would be difficult and expensive, I realized that it would benefit only a favored few. So I dropped the subject for eighteen years. Then in March 1990 came an unexpected breakthrough. While working with a group of Image Streamers, I tried an experiment. First, I gave them a 1-minute summation of Piaget's cognitive schema, as described above. I then asked participants to "let your Image Streaming mind go back to some key point in your intellectual development, whether in childhood or even in early infancy. Your Image Streaming mind knows what that key point is, even if we don't.

The imagers were then asked to describe to their partners whatever arose in their minds, even if it didn't seem to fit with Piaget's schema. Finally, working from whatever image appeared, they were to relax and let that experience move forward into a free-flowing fantasy, in which they were to imagine being enriched by whatever experience or reinforcement was needed.

Without exception, every participant was able to home in precisely on some key developmental episode and to enrich it in their imaginations. They subsequently moved on to other episodes in their lives and repeated the same process until finally arriving at the present.

Future research needs to be done to reveal the extent to which such restorative work affects intelligence, but the uplifting effect upon people using this technique at my seminars has been unmistakable. It is striking how easily and quickly they recall crucial episodes in their development—many of which they had long forgotten. This *hypermnesia*, or enhanced recall, is usually achieved only through profound, and very expensive, hypnotherapy. But Image Streaming provides it free of charge, without the need for hypnosis.

A WORD OF CAUTION!

Memory remains one of the most elusive of all mental phenomena. Psychologist Elizabeth Loftus, for example, has shown experimentally that at least 25 percent of the population can easily be made to "remember" experiences that never occurred.

Quite by accident, Loftus discovered that she herself was among that 25 percent. When she was forty-four years old, an uncle told Loftus that it was she who had discovered her mother's body after her mother drowned in a swimming pool thirty years before. Soon after, "the memories began to drift back," said Loftus, quoted in *Psychology Today*. ". . . My mother, dressed in her nightgown, was floating face down. . . . I started screaming. I remembered the police cars, their lights flashing. For three days, my memory expanded and swelled."

Then Loftus's brother told her that her uncle had been mistaken. It turned out that her Aunt Pearl, not Loftus, had found the body. In fact, Loftus had never seen it.

"I was left with a sense of wonder at the inherent credulity of even my skeptical mind," Loftus wrote.[21]

Below I describe some exercises that will help you remember and enrich peak learning experiences. But I must caution you that it is impossible to say whether the memories that arise are real, partially real, or entirely symbolic and metaphorical. What is important is that the memory means something to you and that your unconscious has presented it in a useful and workable format.

IMAGE STREAMING IS NONHYPNOTIC

Although a trained hypnotist, I long ago rejected hypnotism as an aid in learning and development, not only because it is costly but also because it places too much power in the hypnotist's hands and is far too easy to abuse. Indeed, I specifically developed Image Streaming as a *nonhypnotic* alternative. The imager is fully conscious at all times, fully in control, and no more susceptible to suggestion than a person who is daydreaming while walking down the street.

Nevertheless, when we drop the guards of our memories, the effects can be unpredictable. For this reason, I do not recommend unsupervised Image Streaming to people who are delusional or severely disturbed.

CAVEAT MEMENTOR

If a truly painful memory does emerge while you are Image Streaming, I recommend that you discuss it with a licensed therapist. Even then, you must choose your therapist wisely. In the last few years, millions of people have, with the help of regressive hypnosis, succeeded in remembering traumatic experiences that include not only childhood abuse but also satanic torture, alien abduction, and past reincarnation. I would not presume to say which or how many of these experiences really happened, but Loftus's studies indicate that at least 25 percent may be false memories (and possibly more, since such therapies likely attract a more suggestible personality type).

One respected educator was divorced a few years ago after a therapist encouraged his wife in the belief that her husband had molested her as a child during a past reincarnation. It is not for me to say whether this memory was real or even possible, but divorce seems an overly strong reaction considering the nature of the evidence.

In short, if you must seek therapy, find a practitioner with a responsible attitude toward the use and limitations of recovered memory. A therapist who encourages you to trust blindly in the veracity of such memories is not doing you a service.

It is important to keep in mind that, when using the procedure of Cognitive Structural Enhancement described below, we are seeking not actual memories but rather a key or guide to your past development, however metaphorical or symbolic. Such insights are just as often accompanied by real and recognizable memories as not.

COGNITIVE STRUCTURAL ENHANCEMENT (CSE)

Let us now proceed to the technique of amplifying feedback. Our goal, as explained above, is to retrieve and enrich our memories of what Abraham Maslow called "peak learning experiences." The Image Streaming procedure I use is called Cognitive Structural Enhancement (CSE). Do not attempt it until you have mastered the basics of Image Streaming through several hours of practice.

1. Let your Image Streaming faculties take you back to some key point early in your intellectual development.
2. Describe, to your partner or a tape recorder, whatever comes up in your vision. Let the images unfold freely.
3. Let your Image Streaming faculties imagine for you the most effective and appropriate experience that would have enriched your understanding at that key point. Describe your way through that experience, out loud, to your tape recorder or listener.
4. Let your Image Streaming faculties move you forward in time from that early childhood experience to other key points of development that followed directly from it.
5. Enrich each of these memories in turn by describing them at length in concrete, sensory detail.
6. Continue the procedure until you have worked your way up to the present time.

7. Invest at least a few minutes in detailing, to a partner or tape recorder, the specific improvements that you feel this series of enrichment exercises has made in your present-day understanding and perception.

INSTANT REPLAY

There are various ways to use Image Streaming in working with memories. CSE is only one of them. Image Streaming can also help you review and draw extra content from classes, lectures, conversations, or almost any other type of experience, whether it took place years ago or in the last 5 minutes.

The technique I use for this purpose is called Instant Replay, after the practice of sportscasters in reviewing TV footage of athletic feats. Instant Replay was first developed and brought to my attention by the late Dr. Raymond H. Cameron, Ph.D.

I have adapted and applied Dr. Cameron's technique in accordance with the theory and practice of Image Streaming. While I believe I have improved and expanded its usefulness, I must also take the blame for any shortcomings in the technique as presented here.

Select an Experience

In Instant Replay, unlike in CSE, you select the experience to be reviewed. Perhaps you simply want to gain insight into a conversation you had with another person. Perhaps you attended a class or workshop recently and suspect that it contained more value than you consciously received. Perhaps you had an all-too-fleeting brush with a great teacher, for example, a famous person in your field, and you want to savor and expand on that moment. Or perhaps you are an athlete and wish to review your last game or meet.

Instant Replay can help you in these and many other situations. You can extract the content you need from an experience as conveniently as if you possessed a videotape of the event, available for review at your pleasure.

As you grow adept at Instant Replay, you will see almost every replayed event, however simple or ordinary, grow rich with startling and unsuspected meaning. When you replay a lecture or workshop, for example, you will draw out content of which even the teacher or lecturer wasn't aware.

Instant Replay simulates the manner in which a truly comprehensive genius perceives the world, enfolding the most ordinary situations with layers of meaning, metaphor, and profundity. Replaying your experiences through the Image Stream is like beholding a gorgeous sunset in full color after spending your life color-blind.

Make Lemonade from Lemons

As with CSE, not all the experiences you select for Instant Replay will be pleasant ones. This technique offers the chance to make lemonade from lemons, turning even the most embarrassing, discouraging, or enraging episodes into object lessons to be savored, digested, and transformed in your imagination.

Preparation for Instant Replay

Perform your Instant Replay as soon as possible after the event or situation you have selected. Of course, "as soon as possible" can mean several years after the fact if you have selected a long-ago event. Use a tape recorder or partner. If you use a partner, it is better—but not essential—to recruit someone who experienced the same event.

When you practice Instant Replay for the first time, it is best to select peak learning experiences with obvious special meaning to you. Your success in replaying such experiences will reinforce your Instant Replay responses, and make it easier for you to draw meaning later from more ordinary events, whose lessons have eluded your conscious notice.

THE INSTANT REPLAY PROCEDURE

Think of yourself as an astronaut who has just returned from some exotic mission. You must now engage in a thorough debriefing, describing every detail of your journey to scientists for study back in Mission Control.

1. Close your eyes and practice Velvety-Smooth Breathing.
2. Bring forth a concrete memory of the event you have selected.
3. Describe the event sequentially to your partner or tape recorder, as it really happened, in accordance with the normal rules of Image Streaming. That is, describe it in the present tense and in rich sensory detail. Your partner should keep coaching to an absolute minimum.
4. For the first 4 to 5 minutes, confine your description to sensory impressions of the setting, the teacher or other key person, the other participants, the action, your feelings, and even body awareness. As in normal Image Streaming, do not hesitate to fudge and make things up if it will keep the stream going.
5. When you have established the sensory setting very clearly, begin to open yourself to more abstract insights. Feel free to make comments and judgments about the action before you. Don't worry if these judgments seem wise or correct. If they have popped into your mind, they are important.
6. Soon the images may change or twist in some unexpected, even bizarre way. They may be replaced with entirely new images that at first seem unrelated. These new or changed images are a reflexive response from the broader, unconscious regions of your mind. They are a symbolic representation of some vital insight about the replayed scene. Contained within these images is a mes-

sage about your selected experience. If you persist in your Image Streaming, it will burst upon you in the sort of sudden, all-encompassing insight that Abraham Maslow called an "Aha!" moment.

7. Continue describing the new image for 10 to 30 minutes, or until you achieve a clear "Aha!" experience.

The Best Ideas Come Last

In the creativity technique known as brainstorming, people are urged to throw out ideas freely, without worrying about whether the ideas are good. Generally, the best ideas come at the end of brainstorming sessions, after participants have loosened up and allowed their subtler perceptions to come into play.

In like manner, Instant Replay improves as the session progresses. The last ideas you discover tend to be the most valuable.

15 to 30 Minutes per Day

If you practice Instant Replay for 15 to 30 minutes per day, it will, within 10 to 21 days, exert a profound impact on your life and powers of perception. For thousands of years, men and women have yearned for a second chance to right past mistakes and to draw more value from those rare peak moments that invariably arise too seldom and pass too quickly. Instant Replay offers something very close to that opportunity.

As we have seen, human development arises from a feedback loop—a never-ending progression of expression and response, of trial and error. Image Streaming offers you the opportunity to run that feedback loop again and again until you get it right.

CHAPTER 5

THE "SURPRISE!" EFFECT

fter a long day spent working on his chemistry textbook, Friedrich August Kekulé was frustrated. "It did not go well," the German chemist later recounted. "My spirit was with other things."

Kekulé moved his chair over to the fireplace and stared into the leaping flames. For a long time, he brooded over the benzene molecule, whose exact structure continued to elude him. At last, as Kekulé recalls, he "sank into a half-sleep." What happened next has gone down as one of the great moments—and great mysteries—in the folklore of science.

As Kekulé nodded in his "half-sleep," he saw fantastic shapes and forms in the fire. "The atoms flitted before my eyes," he wrote. "Long rows . . . all in movement, wriggling and turning like snakes." Suddenly, Kekulé was startled by an abrupt movement. "And see, what was that? One of the snakes seized its own tail and the image whirled scornfully before my eyes. As though from a flash of lightning I awoke."

Kekulé realized, in that moment, that his unconscious had given him the key to the structure of the benzene molecule. He spent the rest of the night working out the problem. Soon thereafter, in 1865, he announced the benzene molecule to be a closed, hexagonal ring formed of six carbon atoms—similar in form to the snake in his vision.[1]

"WHAT WAS THAT?"

Visions like Kekulé's are common in science. But whence do they arise? Proper scientific method seems to require that new theories creep up on scientists gradually, through a steady accretion of analysis, deduction, trial and error. In reality, they frequently seem to burst out of nowhere with the suddenness of summer lightning.

"And see, what was that?" exclaimed Kekulé, so startled was he to see a snake seizing its own tail as he stared into the fire. The unexpected quality of Kekulé's epiphany is the best indicator of its brilliance. Only by stealing upon us unawares can ingenious thought break through the censorship of the internal Editor. If an image is not surprising, it is probably not brilliant.

I call this the "Surprise!" Effect.

SQUELCH THE SQUELCHER

In previous chapters, we introduced the concept of The Squelcher, that mechanism in the human brain that tends to shut down ingenious thought. Another word for it is The Editor.

The Editor is mainly a left-brain function. Its natural mode of expression is language, logic, and analysis. Yet it is from the *right* hemisphere of the brain that ingenious insight springs. Confronted by an ingenious thought, the left brain tends to judge and censor it according to conventional wisdom. It squelches the insight with such questions as "Does this idea make sense?" or "Is there any precedent for this?" or even "What will my colleagues think if I express such a view?"

We must therefore learn to bypass or short-circuit the critical left brain. We must in essence "squelch The Squelcher" and allow right-brain insights to flow directly into our consciousness.

Left Brain, Right Brain

I should say a word here about terminology. In general, we say that the left brain deals with language, logic, and analysis as opposed to the right-brain functions of imagery, insight, and pattern recognition. Most researchers today recognize that the left and right hemispheres in fact share many functions. Experiments show, for example, that the visual right hemisphere cannot form complex images without input from the left hemisphere. We will continue to use these terms, however, as roughly accurate metaphors for those opposing parts of our brain that create and judge.[2,3]

Needless to say, ideas do need to be judged and evaluated if they are to be of practical use. As we discuss more fully in Chapter 14, creation and analysis are really two sides of the same process. They must, however, be used in sequence, one after the other. We cannot apply both at once any more than we can mix oil and water. In Chapter 14, we will discuss techniques for focusing tightly on logical, left-brain problems. The "Surprise!" Effect, described here, serves the opposite purpose—it keeps our creative moments dedicated to pure creativity.

CREATE A "SURPRISE!" SPACE

One of the best ways to squelch The Squelcher is to create a "Surprise!" Space in our minds. This is an empty, psychological space that we set aside to attract surprising messages from the right brain, much as we set out a feeder to attract birds.

Kekulé created his "Surprise!" Space by drifting off into semiconsciousness. No sooner had his "Surprise!" Space been cleared, than—Boom!—Kekulé got a surprise.

Virtually all the imaging procedures described in this book are designed to help you clear a "Surprise!" Space in your mind, a vast, empty well into which new perceptions and understandings can pour.

The Problem with Guided Imagery

In recent years, directed or guided imagery has become a popular tool of teachers, athletic coaches, corporate trainers, and even doctors. It resembles Image Streaming in that users close their eyes and view a procession of mental images. But in directed imagery a leader or guide tells you exactly what images to form. A doctor might tell a cancer patient, for example, to visualize white blood cells attacking and devouring cancer cells in some specific part of the body. A trainer might instruct sales representatives to envision themselves closing a big sale. A coach might tell a gymnast to imagine herself executing a perfect routine.

Guided imagery has proved its effectiveness in mobilizing people's talents, confidence, and emotions and even in triggering their immune systems to fight disease. But it is nearly useless in solving problems creatively. The one thing guided imagery lacks is unpredictability. Unless you build a "Surprise!" Space, you cannot hope to get a surprise.

Brainstorming

In 1938, advertising mogul Alex F. Osborn instituted a new technique at his company that his employees quickly dubbed "brainstorming." During brainstorm sessions, employees were encouraged to give voice to any and all ideas, no matter how silly. No one was allowed to shoot down or criticize anyone else's idea. Every thought was recorded without comment. Ideas were examined critically only at a later session.

Osborn's executives were startled by the wealth of unexpected and original insights that poured out of these sessions. In effect, Osborn had created a group "Surprise!" Space. The technique proved so successful that by the 1950s brainstorming had become a fad in America's corporate boardrooms.

"You Reject Too Soon and Discriminate Too Severely"

Part of Osborn's inspiration came from a letter that Friedrich Schiller wrote to a friend in 1788. This friend had complained of difficulty in coming up with fresh ideas.

"The reason for your complaint lies . . . in the constraint which your intellect imposes upon your imagination," Schiller scolded. ". . . for you reject too soon and discriminate too severely." Schiller insisted that "it hinders the creative work of the mind . . . if the intellect examines too closely the ideas . . . pouring in . . . at the gates."

Instead he suggested that his friend allow ideas to flow in freely, deferring critical judgment for a later time. "In the case of a creative mind, it seems to me, the intellect has withdrawn its watchers from the gates, and the ideas rush in pell-mell, and only then does it review and inspect the multitude."[4]

Schiller was talking about building a "Surprise!" Space. If we ask ourselves whether an ingenious insight is practical or sensible, we have already squelched it, because most brilliant ideas seem crazy on the surface. We should never consider either our own conventional opinions or the opinions of others until after we have thought through a problem completely and allowed our genius free rein.

Alex Osborn called this the Deferment-of-Judgment Principle.

Deferment of Judgment

Many of history's geniuses have carried Deferment of Judgment to what may seem reckless extremes. Abraham Lincoln, for example, completely ignored criticism from the press.

"As a general rule," he said in one 1865 speech, "I abstain from reading the reports of attacks upon myself, wishing not to be provoked by that to which I can not properly offer an answer."[5]

Most scientists will not embark upon a course of research until they have first read everything their colleagues have written on the subject. But Albert Einstein scandalized the scientific world by his failure to search the literature.

"It seems to me . . .," Einstein wrote in the introduction to a 1906 paper, "that what is to follow might already have been partially clarified by other authors. However, in view of the fact that the questions under consideration are treated here from a new point of view, I believed I could dispense with a literature search which would be very troublesome for me."[6]

Physicist C. P. Snow marveled that Einstein's famous 1905 paper introducing his Special Theory of Relativity contained "no references" and quoted "no authority . . . the bizarre conclusions emerge as though with the greatest of ease. . . . It looks as though he had reached the conclusions by pure thought unaided, without listening to the opinions of others."[7]

Inventor Michael Faraday went even further than Einstein. The man whom some have called the greatest experimentalist in history often overrode even his own experimental evidence when it ran counter to his inspiration.

"If he had a theory, he was as stubborn as a mule about it," wrote one contemporary. ". . . Most of his experiments were carried out again and again for years, in spite of failures as clear as any experiment can ever be."[8]

Had Faraday taken a realistic look at his failures, he would have abandoned some of his greatest discoveries on the brink of success.

DON'T DEFER IT, DISABLE IT!

Osborn's Deferment-of-Judgment Principle lies at the heart of virtually every creative problem-solving technique in use today. Unfortunately, it contains a serious fault: It is virtually impossible to defer or suspend our judgment. Conventional thinking steals upon us without our conscious control. The slightest whiff of self-doubt is sufficient to paralyze our minds. Willpower alone will not suffice to banish it.

For this reason, techniques that rely on a willful suspension of judgment will always prove dissatisfying to those of us who have not—like Einstein, Faraday, and Lincoln—already stumbled on an unconscious knack for doing so.

The only sure way to circumvent the Editor is to *disable* it (temporarily, of course). There are a number of easy and practical ways of doing this. Some examples follow.

The Speed Principle

Julius Caesar was one of history's greatest military geniuses. He won his battles through a principle he called *celeritas*—speed. Caesar always managed to catch his enemy off guard, arriving on the battlefield days or weeks before he was expected, even if doing so required a forced march through the snow. When a Gaulish tribe called the Bituriges rebelled against Rome, Caesar descended on them so quickly that he caught the rebellious Gauls "tilling their fields without the least fear."[9]

Japanese accelerative learning pioneer Dr. Makoto Shichida uses a tactic similar to Caesar's to outmaneuver learning blocks. He discovered that a speedy input of data can short-circuit the slow left brain. The left brain processes only one word or phrase at a time. But you can learn a foreign language, read a book, or absorb higher math at hundreds of times that rate. The trick, Shichida discovered, is to feed the data into your brain too fast for your conscious mind to follow it.[10]

The Feedback Factor

When a train rolls slowly out of a station and another train rolls into the station on an adjoining track at an equally slow speed, a passenger sitting at the window of the first train will have the impression that the two trains are rushing past each other at twice their actual speed. In fact, the trains' high speed is no mere illusion. Relative to each other, the trains actually are moving quite fast, as would become all too apparent if they were to collide head-on.

A similar effect takes place during Image Streaming. A flood of multisensory perceptions enters your consciousness from the right brain. At the same time, a river of verbal description leaves your consciousness through your mouth and flows back to your brain through your ears in a continuous feedback loop. When these mighty currents rush past and through one another in the narrow, 126-bit channel of your conscious attention, the result is a raging chaos as turbulent as the plunge pool of Niagara Falls.

By this means, we put the Speed Principle to work without actually having to think at high speed. When Image Streaming, we talk, listen, and generate imagery at a normal pace, yet we experience an overload as great as if we were trying to understand a fast-talking auctioneer. Eighty years of psychological research has confirmed that this sort of mental overload clears the way for ingenious thought.

The Brain's Quest for Order

Eighty years ago, most psychologists believed that people built up perceptions in an orderly way, assembling pieces of sensory input much as a child might stack up a tower using wooden blocks.

In 1912, German psychologist Max Wertheimer asked a penetrating question: Why did film audiences perceive the action in a movie as a series of smooth, lifelike motions?

After all, they were really watching nothing more than a jerky succession of still shots.

Wertheimer reasoned that the illusion of apparent motion in film was the brain's way of adjusting to the inscrutable chaos of thousands of flickering still shots. He saw the brain as a homeostatic mechanism constantly striving for equilibrium in a world of maddening disorder. Whenever the brain was thrown off balance by a flood of confusing perceptions, it regained equilibrium by organizing these perceptions into the simplest *gestalt* or pattern that could explain the available information.[11]

Harmony from Chaos

Virtually every beautiful and melodic sound in a symphony orchestra arises from a phenomenon known as standing waves. These are waves that appear to stand perfectly still in space, although in reality they are formed from the interference pattern of two sets of waves in constant collision.

The colliding waves might be caused by a hammer striking a piano wire, a finger plucking a guitar string, or the air rushing through a clarinet or a pipe organ. These waves bounce from one end of the instrument to the other in violent confusion. But from that dissonant chaos arise elegant standing waves, hovering in space like Faraday's rainbow in the mist. It is these waves that produce the pure tones of music.

If It Ain't Broke, Break It

Years ago, corporate managers used to say, "If it ain't broke, don't fix it." Managers feared ruining a smoothly running department with unnecessary tampering. But the prevailing wisdom has changed. Nowadays, enlightened executives have a new slogan: "If it ain't broke, break it!"

They believe that when you break up the old system it will likely come back together in a superior form. This approach—sometimes called "creative destruction"—mirrors the working of the brain, whose *gestalten* seem to increase in elegance in direct proportion to the disorder they are called upon to resolve.

In this regard, ingenious thought behaves much like standing waves. As our perceptions collide in a feedback loop of rapid-flowing perception, left-brain thinking is broken up, and the violence of that collision gives rise to "standing waves" of insight in startling, elegant new forms (see Figure 5.1).

It is precisely from this Rapid Flow with Feedback effect that Kekulé's vision of the snake emerged. The moment

Figure 5.1 When a rapid thought stream outruns your internal censor, it stirs up a turbulent mixture of random perceptions. But, just as a pipe organ transforms turbulent air into elegant standing waves of pure tones, so a well-tuned brain will coaslece chaotic thoughts into elegant standing waves of fresh insight. (The "context" here represents any problem or object on which your mind has focused.)

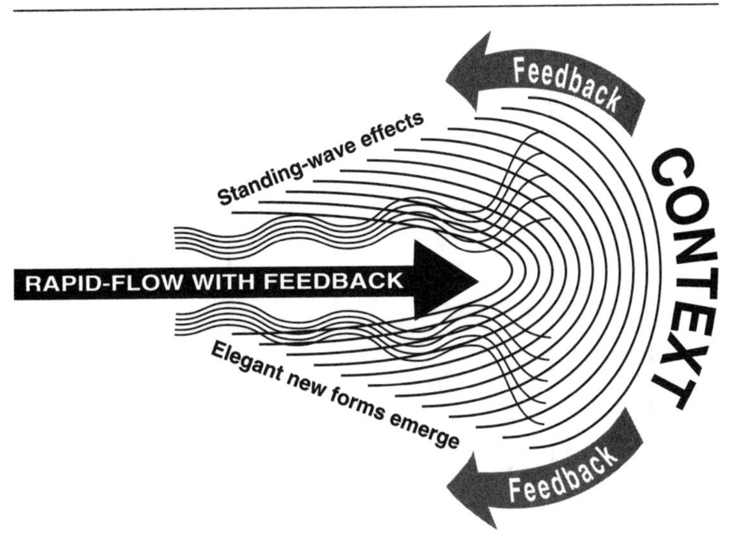

Kekulé thought "What was that?" his opposing currents of perception and introspection had locked into a pattern of standing waves.

CHANGING OUR PERSPECTIVE

Consider the glass box pictured in Figure 5.2. Is the front of the box tilted downward to the left, or is it tilted upward to the right? Stare at it long enough, and the position of the box will seem to flip back and forth between these two options.

You can stare at the cube (called a Necker cube) for quite a few seconds without noticing any change. Then suddenly, without warning, it changes and you perceive an entirely different picture on the page. Yet none of the essential facts have altered. The sensory input is the same. What has changed is your gestalt.

Figure 5.2 Your mind cannot decide how to perceive this cube. Is the front tilted up to the right, or down to the left?

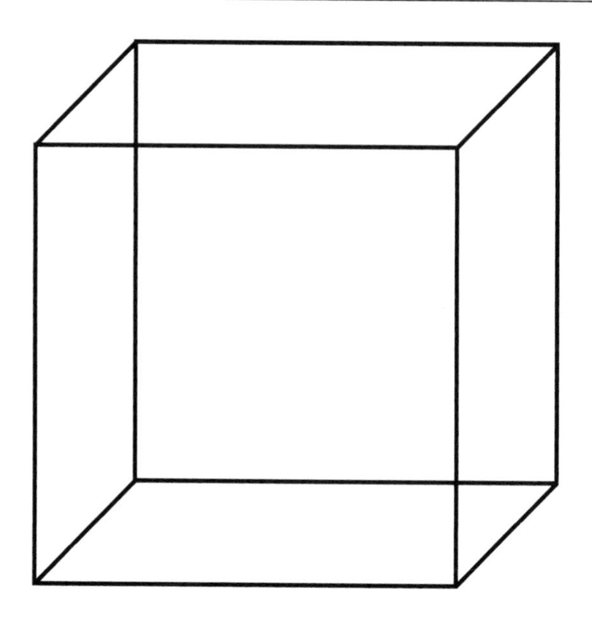

Attractors in the Mind

Now consider a rubber ball tossed into a large, round basin. The ball will roll around in ever smaller spirals until at last it settles on the bottom. Mathematicians would say that the ball is *attracted* to the well of the basin. The point of rest at the very bottom they would call an attractor.

When we flip back and forth between two opposing perspectives on a Necker cube, we could say that each different perspective acts as an attractor for our perception, sucking it alternately one way and then the other.

Now imagine two identical round basins. Each basin represents one of the two different ways of seeing the Necker cube. If we toss a rubber ball into one basin, we would expect it to roll around and finally settle on the bottom. We would *not* expect it to leap suddenly into the other basin, and we most certainly would not expect it to alternately leap back and forth from one basin to the other.

Yet that is exactly what happens when we view the Necker cube. Our perception leaps back and forth from one attractor to the other, never settling on either one. How can this happen?

To make the ball jump from basin to basin, we must apply additional energy from somewhere else. For example, we could strike the bottom of one bowl with our fist, causing a burst of energy that would bounce the ball into the other basin. Then we could strike the other basin, causing yet another burst of energy that sends the ball leaping back into the first basin.

What sort of energy causes our perception to flip back and forth between perceptual attractors?

THE POWER OF CHAOS

Hurricanes are among the most powerful forces in nature. Yet a hurricane can be set off by the tiniest stimulus. Theoretically, a single butterfly flapping its wings on one side of the earth could set off a chain reaction of atmospheric turbulence that would grow into a full-fledged hurricane on the other side of the earth. This is called the Butterfly Effect.

Chaotic systems—such as raging waterfalls, boiling water, churning weather patterns, and runaway stock markets—all have the peculiar ability to leap almost instantly from one state to another upon the tiniest stimulus. One minute, all is calm. The next minute, a tornado has formed. One second, the snow lies quietly on the mountainside. The next second, a thundering avalanche rolls down the slope.

Strange Attractors

In the last twenty years, researchers have concluded that such seemingly chaotic events are really highly organized. Like the rubber ball, such phenomena are sucked back and forth between distinct attractors that can be plotted mathematically. But instead of a single point at the bottom of a basin, their attractors are amazingly complex.

Imagine trying to plot the attractors that account for a pot of water simmering quietly on the stove one instant, boiling over the next, and then settling back to a simmer when the stove is turned down. Plotted on a coordinate grid, such attractors form bizarre patterns of coils and curlicues called fractals.

Fractals have the odd characteristic of being self-imitating. That is, they are composed of smaller and smaller versions of themselves, like the branches of a tree. They also exhibit nearly infinite complexity. The complex and nebulous quality of these figures moved their discoverers to name them "strange attractors."

An Explosive Leap

The incredible flexibility of chaotic attractors has led some researchers over the years to speculate that intelligence may find its basis in the mathematics of Chaos Theory. The Necker cube allows us to derive only two different pictures from the same image, but a mind powered by chaotic attractors could draw an *infinite* number of "ingenious" conclusions from a single set of data.

During the 1980s, two neurophysiologists at the University of California at Berkeley, Walter J. Freeman and Christine Skarda, found strong evidence for this theory. They took electroencephalogram (EEG) readings from the olfactory bulbs of rabbits. The olfactory bulb is that part of the cortex that controls the sense of smell. Freeman and Skarda discovered that when the rabbits inhaled a familiar scent—say, a tasty carrot—their EEG readings showed a burst of extremely regular waves. These bursts exploded across the brain, each wave oscillating in phase with the original carrot pattern, until the entire brain had lit up with the signal for that particular smell.[12]

What surprised Freeman and Skarda was how quickly the rabbits' brains flip-flopped back and forth from a full-fledged "burst state" to a "nonburst" state, with no sign of a gradual transition in between. This phenomenon was a strong indicator of chaotic activity. Freeman and Skarda confirmed their hunch by creating a computer model that made three-dimensional plots of the burst waves across large regions of the rabbits' brains. Sure enough, the computer yielded complex, coiled figures typical of chaotic attractors.

"The images suggest," Freeman concluded, "that an act of perception consists of an explosive leap . . . from the 'basin' of one chaotic attractor to another."[13]

Complexity Is the Key

Freeman and Skarda proposed further that such rapid state change was the very mechanism that enabled human brains to "produce novel activity patterns" and to "generate insight."

Complexity is the key. A butterfly cannot peddle a bicycle by flapping its wings. That system is too simple for the Butterfly Effect to work. But it can theoretically start a hurricane, because the atmosphere is so vast and filled with so many intricate eddies and currents, down to the molecular level, that even the smallest disturbance will affect trillions of other variables through a series of incalculably complex chain reactions.

The more complexity we bring into our thought process, the more room we create for dramatic state changes. Complexity actually widens our "Surprise!" Space.

PHASE RELATIONSHIPS

Any given stimulus enters the brain at a distinct "port of entry." The signal is then processed, tagged, and shunted off to other parts of the brain for further processing. The problem is that once the stimulus has been tagged with labels and instructions at the port of entry, other parts of the brain will tend to follow those first instructions.

I call this the Close-Out Effect. The port of entry has effectively closed out on the stimulus, hanging a label on it that says, "Here's how we handled it, folks." Other parts of the brain are thereby warned not to waste time with further processing.

To build a "Surprise!" Space, you need to get as many different parts of the brain as possible working on that stimulus simultaneously. You want the *phase relationships* in your brain—the sequence by which different brain regions fall into phase with one another—to be timed so closely that any given stimulus is discharged across the brain before close-out instructions can be attached.[14,15]

This is exactly the effect of Image Streaming and other Rapid-Flow-with-Feedback techniques.

Dali's Clocks

Suppose you glance at a clock. After about half a second, your visual cortex announces that it has seen a "clock" and flashes that message to the rest of the brain. The Close-Out Effect takes hold.

At that point, your brain thinks it has only two choices. It can study the clock to see what time it is, or it can ignore the clock. After all, what else can you do with an object labeled "clock"?

But suppose you didn't close out on the clock? Suppose you managed to short-circuit the Close-Out Effect, allowing other parts of your brain to process the clock in an "unclocklike" way? Suppose, for example, that you started noticing the sheen of light across its crystal face, the elegant form of its numerals, the hypnotic tick of its second hand, or the smoothness of its fine metal casing? Suppose you began to reflect on the ephemeral nature of time and the evanescence of matter?

If you persisted long enough, you would likely experience a sudden, ingenious shift in perspective, similar to that of the surrealist painter Salvador Dali, whose famous melting clocks are among the most powerful and haunting images in modern art (see Figure 5.3).

IF IT FEELS WRONG, IT MUST BE RIGHT

Remember how agitated poor Kekulé grew before he drifted off in front of his fireplace? The last thing in the world he expected was a brilliant insight that would solve his daunting problem—and make him famous, to boot. On the contrary, Kekulé felt confused and defeated, unable to work further.

If only Kekulé had known that his despair was actually a good sign! It meant that his left brain felt defeated in

Figure 5.3 When you look at a clock, certain areas of your cortex announce to the rest of your brain, "This is a clock." Other brain areas then suppress further perceptions that don't relate to the practical function of a "clock." In effect, your recognition area has closed out further processing, telling the rest of your brain, "Forget, it folks! We've already handled that problem! It's just a clock." This is called the Close-Out Effect.

However, if other areas of the brain are allowed to keep evaluating the clock, ignoring its "clock" label—as happens when phase relationships are close enough that other brain regions can short-circuit the Close-Out Effect—a mere clock can be transformed into an artistic masterpiece such as Salvador Dali's "The Persistence of Memory."

its efforts to make sense of the surrounding chaos. It was finally ready to give up and get out of the way.

A sense of confusion is the surest sign that your Squelcher has been effectively squelched. When your internal Editor is up and running, you don't feel confused at all. On the contrary, everything seems very clear and simple. Only when an ingenious thought is bubbling in the pot do you feel as agitated and perplexed as Kekulé pondering those swirling serpents in the fire.

Don't expect your "Surprise!" Space to be a comfortable living room. It is an alien landscape filled with turbulence and confusion. That's why it takes a certain daring to be a genius. When despair creeps up on you, drink deeply of it. Savor your moment of confusion. Let it swirl around you like Kekulé's serpents. It means your victory is near.

A TYPICAL "SURPRISE!"

In Richard Poe's Image Stream, described in Chapter 3, the vision of a cheetah on the African savannah followed quite naturally from Richard's initial image of a pattern of spots. The story of Richard and his friendly cheetah then unfolded in a logical manner, almost as if we were viewing the opening scene of a Walt Disney wildlife epic. We might have expected the happy pair to go bounding off in pursuit of a gazelle. But that would have been too logical, too obvious, too *unsurprising*.

Instead, without warning, the sky was ripped asunder by a gigantic Lucrezia Borgia character in fifteenth-century Italian garb who gazed down from a peculiar "heaven" that looked, smelled, and felt like a musty alchemist's laboratory.

This sort of bizarre digression—typical of dreams and Image Streams—represents the intrusion into your mind of a potent, unconscious message. Such intrusions are not always bizarre. They can be as mundane as a red, high-heeled shoe turning up in an odd place or an old automobile tire that just won't go away. Virtually any unexpected occurrence in an Image Stream is sure to have emanated from some unconscious region beyond The Squelcher's reach.

In the next chapter, we learn how to decipher these enigmatic messages. For now, we need only learn how to stimulate and achieve such vital breaks in the Image Stream.

TECHNIQUES FOR INDUCING THE "SURPRISE!" EFFECT

Often, especially when you are just starting out, you may find yourself Image Streaming for a long time with no apparent resolution or obvious "Surprise!" In such cases, you may be forced to meddle a little bit with the Image Stream.

In general, meddling is not a good idea. The more absorbed you become in a stream of imagery, the deeper you fall under the spell of your right brain and the more profound will be your imagery. Any kind of conscious or left-brain meddling tends to break the spell and threaten the spontaneity of your Image Stream.

Nonetheless, I have developed some techniques for artificially inducing the "Surprise!" Effect that will do minimal damage to your spontaneity. Some techniques are detailed below.

Create an Answer Space

Failure to achieve a "Surprise!" generally arises from some sort of inhibition. You can trick your subconscious mind into thinking it's safe to reveal the "Surprise!" by providing an Answer Space that is "hidden from sight". Any opaque barrier will do. You might imagine your Answer Space hidden behind a wall or shut behind a thick dungeon door (see

Figure 5.4). The important thing is that you can't see it. In effect, you are telling your subconscious mind, "If you unveil the 'Surprise!' behind that barrier, I promise not to peek until you're ready."

Figure 5.4 The "Thesholding Technique." Envision a door concealing a mysterious "Answer Space" beyond. When the door opens, the answer to your question is revealed.

Thresholding

Eventually, of course, you will have to peek, but first you must give your shy inner genius time and privacy to put together an image without your kibitzing over its shoulder. In order to peek, you can use a set of techniques called Thresholding. For reasons that will become obvious, the particular Thresholding technique described below is called Over the Wall:

1. Decide what problem or question you want to solve. Then put it aside in some less conscious part of your mind while you turn your attention elsewhere.
2. Set a timer or an alarm clock for 7 to 8 minutes. Try to use a timepiece that gives off a few pleasant "bings" or chimes rather than a strident uproar that will break your concentration and force you to turn it off.
3. Begin Image Streaming, this time with the following directed image.
4. Imagine that you are standing in the midst of a strangely beautiful garden. A great wall stands before you. This wall represents your Threshold. Beyond it lies your Answer Space.
5. Spend several minutes describing the garden in rich sensory detail.
6. Now walk up to the wall and describe it. Put your hand on the wall. Lean your face up against it. Feel the damp, slippery moss on its old stones (or whatever texture you have imagined for it). Describe how it smells.
7. Block out any conscious thought about your question or problem. Focus only on sensory impressions.
8. The instant you hear the alarm go off, jump over the wall! Suddenness is the key. You must catch the answer before your left brain has a chance to distort it with conscious and conventional expectations.
9. Seize upon your *very first impression* of what lies beyond the wall, even if it lasts only an instant. Continue describing whatever you saw, from memory, even after the image

has fled. Your flow of description will soon bring the image back.

In general, as with ordinary Image Streaming, the degree to which you are *surprised* by what you see in the Answer Space is the surest indication that it is a fresh, original insight. It doesn't need to be profound. Indeed, a trivial object might be the very symbol that conveys the answer to your problem. It must simply be unexpected.

If what you see proves unsurprising and conventional, then you probably didn't jump quickly enough or you allowed your mind to plan in advance what you were going to see. In this case, repeat the exercise and try not to anticipate your "Surprise!" until you hear the sound of the alarm.

Thresholding is only a tool. Do not treat this technique as an exercise in guided imagery where you feel obliged to stick with the program, no matter what. In fact, you may start to see a stream of spontaneous imagery at any time before the alarm goes off. If this happens, simply abandon the Thresholding exercise. Shut off the alarm and start describing whatever images you see. They will no doubt lead you to your "Surprise!" much more quickly and directly than Thresholding would.

Debrief

Try this experiment. Form an idea about any subject that is of interest to you. Now write out a single paragraph about that idea on a sheet of paper. Take that same paragraph and type it on a typewriter, if you have one. Now type it into a computer and print it out in various type fonts and point sizes.

When you reread your paragraph in these different media, you will find that even slight differences in context evoke significant differences in your thoughts and perceptions as you feed back your own writing to your mind. In their experiments on rabbits, Freeman and Skarda discovered that the electroencephalographic map of a particular perception can alter radically depending on the context or associations linked to that perception. I call this the Context Principle.

You can put the Context Principle to work in deciphering your surprises. After you finish Thresholding, perform an additional step. Debrief by describing quickly, to another person or to a notepad, whatever you remember about your experience on the far side of the wall (in your Answer Space). The important thing is that you debrief to a *different medium* from the one you used while Image Streaming. In other words, if you spoke originally into a tape recorder, don't debrief into a tape recorder. If you spoke to a partner, don't debrief to that same partner (a fresh partner, however, will do nicely).

Quick Thresholding

You may find it easier to use a technique I call Quick Thresholding. Coauthor Richard Poe used this method in his Image Stream described in Chapter 3. When he stared into the fiery oven, he saw an object arise that appeared first to be a crystal ball and then a pulsating, reptilian egg.

Richard had no way of knowing what the egg contained, but he decided on the spot that whatever came out of that egg would be his answer. It turned out to be a dragon. That's Quick Thresholding.

Quick Thresholding has the advantage of playing off an Image Stream that is already in progress. You are therefore still deeply immersed in the spell of your right brain, and you don't need to spend time and energy reestablishing neurological contact with the imaged scene.

Virtually any opaque object that appears in your Image Stream can serve as a Quick Threshold. It might be a door, a window shade, a bend in the hallway, a thicket, or a photo album cover. If it is physically capable of concealing an Answer Space, it is a suitable Thresholding device.

The Quick Thresholding procedure follows:

1. Begin while your Image Stream is in progress or return to some critical, remembered scene of a previous Image Stream.
2. Select a Thresholding device.
3. Identify the Thresholding device aloud into your tape recorder, and describe it *briefly*, just enough to solidify the object in your imagination.
4. Now decide what question you want to pose to this Threshold. A good question to start with is "What is the meaning or message of this spontaneous Image Stream?"
5. Pose your question softly out loud and in your mind as loudly as you can.
6. While asking the question, reach out and "touch" the Thresholding device. This metaphoric contact "programs" the Threshold.
7. Now widen your neurological contact with the Thresholding device. Begin describing it in earnest, using all five senses.
8. Whenever you are ready, jump or cross over the Threshold abruptly. This could mean opening a door, stepping through a gateway, raising a window shade, or flipping open a photo album, depending on the form of your Thresholding device.
9. Seize upon your very first impression of whatever lies in the Answer Space beyond, and describe it.
10. Step through into the Answer Space and increase your neurological contact with the whole scene, describing it in detail.
11. Once you have fully established your physical sense of the Answer Space, find some object in that space that can serve as a new Thresholding device. Each time you repeat the process, you will be more deeply engrossed in the spell of your right brain and closer to your subtlest and most ingenious perceptions.

WIDENING YOUR DOORWAY

If you have faithfully practiced the techniques outlined in these first five chapters, you know from experience that Image Streaming becomes dramatically easier the more you do it. Each time you pass through the doorway that leads to your Image Stream, the door grows wider and taller. Soon you will be able to slip back and forth without a second thought.

Ultimately, you will learn to consult your greater resources so easily and quickly that you will be able to do so as quickly as an eye blink in the middle of a conversation, with no one else noticing. Quick Thresholding is especially useful in building this skill, because it teaches you to interact with and carry on a conscious dialogue with your right brain.

When you have reached this level of proficiency, you will be an intuitive giant capable of summoning startlingly acute insights at will, in every situation, throughout the course of your day.

CHAPTER 6

INTERPRETING THE IMAGES

King Croesus of Lydia once consulted the oracle of Delphi to see whether he should march against the Persian Empire. "If you do," promised the oracle, "you will destroy a mighty empire."

Croesus was elated. What empire could be mightier than Persia? The oracle had virtually guaranteed that Croesus would destroy it. Filled with confidence, the Lydian king marched . . . only to be crushed by the Persians! In his rush to seize upon an encouraging interpretation, Croesus had neglected to ask *which* empire the oracle was referring to. It was his own Kingdom of Lydia that fell to ruin.

THE THREE PITFALLS OF INTERPRETATION

Interpreting a Delphic oracle is not so different from deciphering an Image Stream. Three common pitfalls prevent most people from interpreting their imagery correctly. King Croesus fell headlong into all three. They are:

1. Predetermined expectations of what the answer ought to be.
2. Settling for generalities instead of nailing down the particulars.
3. Lack of persistent follow-through.

Croesus was already eager for war, which meant that he probably would have heard a positive message no matter what the oracle said. So much for pitfall 1. In promising that Croesus would "destroy a mighty empire," the oracle uttered a useless generalization. Croesus should have noted that her words were ambiguous. Instead, he accepted them at face value, plunging right into pitfall 2.

Pitfall 3 was probably the one that did him in. A little follow-through on Croesus' part might have changed world history. All he had to do was ask, "Which empire do you mean?"

EIGHT STEPS TO A SOUND INTERPRETATION

Interpretation, like Image Streaming itself, grows easier with practice. Eventually, you will gain an instinctive feel for the language of your right brain, enabling you to make snap interpretations much of the time. When you're just beginning, however, you should adhere to the following eight-step regimen:

1. Decide if the message is literal or symbolic.
2. Distinguish fact from feeling.
3. Identify key associations.
4. Apply your personal decoder.
5. Apply the "When-Then" Test.
6. Last is best.
7. Use Thresholding and Questioning techniques to nail down the particulars.
8. Synthesize an "Aha!" experience.

"Thou Shalt Interpret Thine Own Images"

The first commandment of interpretation is "Thou shalt interpret thine own images." Its corollary is "Thou shalt not interpret for thy partner."

There is nothing wrong with getting a fresh perspective. Any outside opinion, whether from a friend, a spouse, a therapist, or even one of the popular dream books on the market, can help you think through your imagery. However, you should avoid developing a dependency on an outside soothsayer. Such a subservient relationship not only stunts your growth as a self-interpreter, it also makes you vulnerable to manipulation—deliberate or otherwise—from others.

Especially in the early stages, it is far more important for you to build up your interpretation "muscles" than to decipher a particular image. It is better to draw your own wrong conclusions than to take the shortcut of taking someone else's interpretation, even if it does turn out to be right.

When working with a partner, of course, you will share your images constantly. However, your partner should refrain from imposing his or her interpretations on your imagery, and vice versa. Even when working alone, most people sooner or later will want to share their images with others. The best rule of thumb when working solo is to work through all eight steps *on your own* before discussing your Image Stream with anyone else.

Step 1: Literal or Symbolic?

Sometimes images are completely literal in their meaning. When Bob S. in Chapter 1 saw the worn-out tire from his fiancée's car, it meant nothing more esoteric than, well, the worn-out tire from his fiancée's car!

For obvious reasons, literal images are the easiest to decipher. Unfortunately, their literalness is not always immediately apparent. It took Bob S. a few minutes to realize where he had seen that tire before.

Psychologists have learned that it is easier to gain access to suppressed or subliminal information when you are relaxed. Try doing some Velvety-Smooth Breathing, as described in Chapter 3. When you are relaxed, just think about the image for a while and see if something pops to mind. If it doesn't, your image is probably symbolic. Proceed to the next step.

Step 2: Fact Versus Feeling

In analyzing dreams, Jungian psychologists distinguish sharply between observation of actual events in the dream and suppositions that we impose after the fact. This technique is useful in cutting through our conscious distortions and getting to the bedrock content.

In Image Streaming, too, facts are more important and reliable than whatever feelings or impressions we may have about those facts. In the transcription of coauthor Richard Poe's Image Stream in Chapter 3, for example, we find the following passage:

> I can look out and there's a waterhole out there with zebra and wildebeest and a bright sunlit scene, a grassland that stretches out over the horizon. The wind is very hot against my skin, and I feel at one with the cheetah. We're together and we belong together, and I'm not a stranger in this place . . .

The first half of this passage consists entirely of *fact*. We see physical descriptions and nothing more. But then Richard claims that he feels "at one with the cheetah" and extrapolates from there. Richard here is conveying not bedrock content but his own *feelings* or impressions about what is going on, feelings that may or may not be true, according to the original intent of the Image Stream. Note that earlier in the same Image Stream Richard said:

> . . . I can see the cheetah's head. . . . It turns to me and I can touch its head. I can feel its ears, and their floppiness, the rubbery floppiness and feel the drool of its mouth as I stroke it all over its head and it looks at me, regards me

> warily, not warily but in an accepting way, as if I'm part of its cheetah family . . .

Richard's very first impression is that the cheetah eyes him "warily." Only afterward does Richard change his mind and decide that it eyed him "in an accepting way." Or perhaps it was first wary and then *became* accepting. We really can't tell from the transcript. Richard has assured me that, looking back on the experience, he remembers seeing no clear indication on the cheetah's face of *any* particular emotion. Its feline features were quite inscrutable to him. So his conscious mind filled in the blanks by offering two entirely opposing interpretations, one after the other.

Although these feelings or impressions that Richard had about the cheetah's state of mind are still useful in interpreting the Image Stream, they must be regarded as a *secondary* reaction, a sediment laid over the bedrock images. We might think of the concrete images as a metaphorical Torah—the inspired Word of God—and Richard's secondary feelings about the images as merely a Talmud—a collection of rabbinical commentaries, which may or may not be totally on the mark. Christians would recognize the same difference between Scripture and sermon.

Like a rabbi probing theological issues, we must ground our conclusions firmly in the Torah, the bedrock imagery. Talmudic feelings can fire our speculations, but we must back up each interpretation with Scriptural fact.

Paradoxically, knowing where the bedrock lies gives us greater freedom to explore our secondary feelings and associations. Taken alone, such speculations could easily lead us down the garden path, but with a firm grounding in fact we can find our way home, no matter how far we roam.

Step 3: Identify Key Associations

Associations are simply those secondary thoughts that the images bring to mind. Think back over your Image Stream (or listen to the tape, if you made one). Each image will evoke some clear association as you think about it. Write down that association as it comes to mind, even if it seems silly.

Richard says that the cheetah scene made him think of an episode he had seen from the PBS "Nature" series, about a mother cheetah raising her cubs on the African savannah. The Lucrezia Borgia character in Italian Renaissance garb put him in mind of my "Project Renaissance"—the institution through which I disseminate and promote my accelerative learning techniques. The startling manner in which "Lucrezia" ripped open the sky suggested to Richard a traditional image of the Christian Apocalypse, while the alchemist's laboratory in Lucrezia's "heaven" seemed to Richard to be an artifact of his own lifelong interest in mystical arcana.

The dragon image evoked for Richard his love for ancient and medieval folklore. Its black color suggested to him both power and evil. As for the dragon's sudden departure into space, Richard notes that the creature left the solar system in a leftward direction, at an angle of about "10 o'clock," taking an imaginary course that Richard has linked since childhood with the route to Alpha Centauri, the setting of a science fiction novel he enjoyed as a boy. He also associated the space scene with Einstein's Theory of Relativity and with hard science in general.

Because these were the clearest, strongest, and most obvious associations that Richard could discern, we will regard them as the key associations in his Image Stream.

Step 4: Your Personal Decoder

"I have made it a rule," wrote Carl Jung shortly before his death in 1961, "to remind myself that I can never understand somebody else's dream well enough to interpret it correctly." [1]

As a protégé of Sigmund Freud, Jung had chafed under the great man's dogmatic approach to dream analysis. Freud believed that whenever cigars, towers, flagpoles, battering rams, or virtually any other cylindrical object appeared in a dream, it always meant you-know-what. Jung thought that was silly. He thought that symbols meant different things to different people, depending upon the context.

"Learn as much as you can about symbolism," Jung advised his students, "then forget it all when you are analyzing a dream." [2]

I agree with Jung. Fifteen years of working with Image Streamers has convinced me that each imager has a personal code of symbolism that is different from every other imager's. Only you can accurately decode your own symbols.

Keep a Codebook If you are serious about Image Streaming, it would be well worth your time to start keeping a codebook, listing all the objects, people, and situations that seem to recur in your Image Streams together with what you think they might mean. Often, there will be more than one possible interpretation, especially at first. As you become more experienced with Image Streaming, you will gain an ever-clearer perception of what various images mean to *you*.

Step 5: The "When-Then" Test

Another technique borrowed from Jungian analysis is one that I call the "When-Then" Test. It is useful for exploring the narrative *sequence* of events during an Image Stream.

In Richard's Image Stream, a crystal ball emerged from the blazing athanor and turned into an egg. A dragon emerged from the egg, flapped its wings, and flew off into space. Richard then jumped on its back and rode the creature out into the interstellar void.

In and of themselves, such images can tell us much, but additional meaning is embedded in their sequence. Why did these events happen in the particular order they did?

To probe this question, apply the "When-Then" Test. We might say, "*When* the crystal ball emerges from the athanor, it *then* becomes an egg." Reframing the action in this way helps sensitize us to hidden cause-and-effect relationships. For example, does the crystal ball *always* become an egg when it is removed from the fire? If we put the egg back into the fire, will it turn back into a crystal ball? What quality does the fire have that inhibits or prevents the egg from unleashing its egglike nature?

Other "When-Then" relationships in Richard's Image Stream might include:

- "*When* a dragon emerges from the egg, *then* it flies off into space."
- "*When* the dragon flies off into space, *then* Richard jumps on its back."
- "*When* Richard feels a oneness with the cheetah, *then* Lucrezia Borgia rips open the sky."
- "*When* Richard smells Lucrezia's perfume, *then* he is whisked into a celestial alchemist's laboratory."

Step 6: Last Is Best

The longer you persist in a particular Image Stream, the more deeply engrossed you become in the spell of your right brain. For that reason, just as in brainstorming, the ideas and images that appear *last* in your Image Stream are the ones most likely to be free of conscious distortion and inhibition.

There is plenty to be learned from studying the Image Stream as a whole, following its narrative transitions from scene to scene. But your efforts will bear the greatest (and quickest) fruit if you focus on the last few moments of the experience.

Step 7: Nail Down the Particulars

If you come up with an interpretation that is profound and far-reaching, it is probably too *general*. Wide-ranging philosophical principles are of little use in problem solving. They are also easy to fudge, as carnival fortune-tellers know all too well. Any statement that is sufficiently vague and general will always seem to fit the facts.

It wasn't hard for Richard to come up with a vague sense of his Image Stream's underlying theme. In fact, his wife, Marie, suggested it to him immediately when he described the Image Stream to her (yes, I'm afraid my coauthor cheated).

Marie suggested that the dragon symbolized Richard's interest in all things magical, mythical, and ancient. The fact that Richard rode the dragon into space therefore represented a merging or harmonizing of Richard's passion for archaic esoterica with his equally strong interest in modern science.

Marie's interpretation seemed to Richard to be right on the mark. But it was too general. It did little to solve the *particular* problem Richard was facing at that time—how to meet an impossibly short deadline for completing *The Einstein Factor*. Somewhere embedded in that Image Stream lay the answer, but it would not yield its secrets without persistent *follow-through*.

Follow-Through via Thresholding Techniques If the Image Stream offers an ambiguous or unsatisfactory message, Thresholding will enable you to probe more deeply. Following the principle of "last is best," Richard looked at the final sequence of his Image Stream and selected the pulsating, reptilian egg as a Thresholding device.

In Richard's original Image Stream, the egg had burst open to reveal a small, black dragon. Reentering his Image Stream, Richard asked the egg to give him an alternate image. On the first attempt, the egg once more yielded a small, black dragon, identical to the previous one. Fearing that the dragon had appeared in response to his conscious expectation, Richard repressed the image (another no-no on the part of my coauthor, I might add: nobody's perfect) and found himself looking at a Mona Lisa, smiling enigmatically from the darkness.

Unsatisfied, Richard continued Thresholding. His next attempt yielded a goldfish in a bowl, swimming around a tiny white effigy of an Olmec head. Richard continued to image freely and found himself involved in scenes from Cecil B. DeMille's *The Ten Commandments*. He even stumbled upon Charlton Heston as Moses, presiding over the first Passover supper. Richard then asked Moses to impart to him the secret of the egg, at which point Richard suddenly saw a bevy of comic, animated cartoon characters dressed as Greeks of the fifth century B.C. scuttling around, Charlie-Chaplin style, at the mouth of a deep cave or tunnel, which seemed to symbolize Plato's famous metaphor of the cave in *The Republic*. Richard plunged into that tunnel, going deep into the earth, expecting to find an answer in the caverns below. Alas! He found only a Cyclops holding a club, wearing a great shaggy skin and "looking somewhat harmless and forlorn."

Find the Common Elements Most people won't have the time, patience, or ability to painstakingly decipher every image and "when-then" relationship in an Image Stream, especially when the number of Thresholding variations starts to pile up as it did for Richard. A shortcut to interpretation is to ignore those aspects of the imagery that seem obscure and search for common elements among all the various versions that seem to sum up the meaning of the whole.

As Richard reviewed the various Thresholding alternatives to his dragon image, it occurred to him that all the alternatives had a joking or mocking quality, as if they were taunting him for trying to find something that wasn't there. The Mona Lisa figure smiled at him enigmatically, as Mona Lisas are wont to do, refusing to give up her secret. The goldfish that emerged from the athanor seemed a poor substitute for the real gold that alchemists normally cook up in their ovens. Seeking to glean the Word of God from the lips of Moses himself, Richard was dismayed to see only cartoon characters of Greeks scurrying to and fro, Charlie-Chaplin style, as if in deliberate mockery of the profound philosophical gravity of Plato's cave, looming beyond. Finally, after plunging into the cave in an effort to plumb the darkest depths of the subconscious, Richard came face to face with a forlorn and pathetic Cyclops, whom he says was modeled after a character in a silly TV series about Hercules.

Some additional clarification techniques that Richard could have used, but didn't, can be found later in this chapter.

Step 8: "Aha!" Experiences

All of this suggested to Richard that he was looking in the wrong direction. Instead of finding alternatives to the dragon, he should look back at the dragon itself, which, after all, had sprung from the egg in two consecutive versions.

First "Aha!" That was Richard's first "Aha!" The dragon—and *only* the dragon—really did represent the answer to his question. There was no alchemist's gold, no Word of God, no Platonic profundity or Underworld revelation to provide a shortcut. For some reason, Richard's Image Stream required him to confront the dragon *as* a dragon.

Second "Aha!" Richard got his second "Aha!" by cheating. In flagrant violation of the first commandment of interpretation, he broke down and sought advice from his wife, Marie. The two of them discussed Richard's Image Streams at length, tossing suggestions back and forth. At one point, their conversation took a turn something like this:

> MARIE: What does a dragon mean to you?
>
> RICHARD: It means wisdom. It is like a snake, which imparts knowledge, but can also be dangerous and evil, like

> the snake that deceived Adam and Eve and caused them to eat from the Tree of Knowledge. So it represents both the good and bad possibilities of knowledge.

Richard was applying his personal decoder, sharing with Marie his own idea of what a dragon meant gleaned from his extensive readings in mythology and folklore. For other people, a dragon might mean something entirely different.

Armed with this information, Marie realized that the alchemist's lab, like the dragon, also represented arcane wisdom with an edge of danger or evil. The hot, musty alchemist's lab was, moreover, confining, like a prison. It was a realm governed by ancient, esoteric traditions dating back thousands of years, all the way back to ancient Egypt according to some traditions. By leading Richard away from that musty environment, the dragon was helping to liberate him.

Richard's ride on the dragon was thus a reckless but courageous maneuver, enabling him to harness a dangerous but liberating force that could lead him into new and better worlds or, more particularly, into a starry cosmos that—in Richard's personal codebook—represented the serene realm of pure scientific inquiry.

Marie suggested that Richard should "let go," "break all the rules," and not worry about what critics or other commentators would say about *The Einstein Factor*. Strangely enough, this advice struck home on a very *particular* issue that had been troubling Richard concerning the writing of Chapter 7—the very next chapter he was planning to write!

DEEP DECODING

The blocks and inhibitions that hamper our work usually arise from deep-rooted conflicts that persist in our minds over many years. Through the Image Stream, we can identify these inner demons and begin to exorcise them. Such Deep Decoding often sheds unexpected light on the specific meaning of an Image Stream.

When Marie told Richard to let go and break all the rules, he recognized instantly the particular rules that were holding him back. As a boy, Richard had been enthralled by science, spending his happiest hours observing pond scum under a microscope. By the time he was twelve, Richard could stain a blood slide and perform such clinical tests as differential white cell counts, having been taught these skills by his microbiologist mother. Richard was fifteen when he won a National Science Foundation scholarship to study geology for a summer at Syracuse University. He subsequently matriculated at SU at the age of sixteen, as a premed freshman. In short, Richard was thoroughly grounded, from an early age, in the rules of science.

During his freshman year at college, however, Richard underwent a dramatic change that shocked and appalled his teachers and parents alike. He suddenly lost interest in hard science and switched his major to creative writing. He embraced all things mystical and esoteric, keeping Jungian dream journals, immersing himself in novels by Hermann Hesse and Jack Kerouac, and pondering the countercultural philosophies of Tom Wolfe's *The Electric Kool-Aid Acid Test* and the *Tibetan Book of the Dead*. He later dropped out of his masters' degree program to study poetry with Allen Ginsberg at a Buddhist school in Colorado and then trekked to California to write a novel. For the next twenty years, Richard pursued a writing career, abandoning all plans to become a doctor or a scientist.

Richard later came to regard his detour from science as a failure. When he began coauthoring *The Einstein Factor* with me, Richard looked upon the project as a chance for redemption, a way to restore his credibility as a scientific thinker. He therefore outdid himself in researching the scientific background of my theories, striving to present each point with exacting technical detail.

RIDE THE DRAGON!

Then came Chapter 7. As is perhaps inevitable in a speculative field like accelerative learning, my research has at times carried me to that far edge of human experience where the familiar explanations of science falter. These areas are treated most bluntly in Chapter 7.

For Richard, as my coauthor, they presented a problem. Would a forthright treatment of the spookier aspects of accelerative learning dampen the book's scientific credibility? Should the book skirt these phenomena or rationalize them with flimsy but conventional explanations? This would certainly be the safest course. Yet it would also sweep some of the key discoveries of Project Renaissance under the proverbial rug. Until his "Aha!" experience with the Image Stream, Richard didn't realize how profoundly this dilemma troubled him.

What finally broke the ice was the image of the dragon riding into space. The dragon represented all that was dangerous and mysterious in Chapter 7. It grew in size and flew from the earth, becoming ever wilder and less manageable. Yet, in the end, Richard not only gained control of the dragon but actually harnessed it as a vehicle to "ride off" into the Einsteinian cosmos.

Here was a clear promise from Richard's unconscious that, if he would cast aside his fears and boldly confront the tough issues in Chapter 7, he would "ride the dragon" to a happy but honest synthesis of science and speculation.

BREAK THE RULES

Of course, as we observed above, Richard cheated by discussing the Image Stream with his wife. When he sheepishly pointed this out to Marie, she countered, "So what? Break the rules!"

This, too, was in keeping with the image of riding the dragon, and in general it's a good principle for all Image Streamers to keep in mind. No rules are so sacred that they should not be broken from time to time. Indeed, I have discovered that it is sometimes quite useful to discuss my images with other people, especially those, like my own wife, who know me and my personal code very well. Seek first to work out the meaning on your own. But when you get stuck, by all means seek the advice of others. Be wary, however, of accepting their advice out of subconscious politeness. Be especially wary of accepting answers simply because they were imparted by someone with more authority or more formal credentials than yourself.

THE PRIORITY PRINCIPLE

Your Image Stream knows better than you which issues are really important. Often it ignores your overt question and answers an entirely different question that perhaps never consciously occurred to you. This is what happened to Bob S. in the first chapter, who found himself staring at an enigmatic automobile tire. In general, your Image Stream will prioritize whatever issue or problem is most urgent at the time. I call this the Priority Principle.

Richard entered his Image Stream with the conscious assumption that his most important problem was his short deadline. He therefore probed each image for some hint as to how he might speed up the writing process or cut down the number of chapters. Instead, his subconscious mind advised him on an entirely different subject. The advice it gave probably *added* to Richard's work load rather than relieving it. Nevertheless, it resolved for him the one problem that proved most critical in writing the book.

WHAT IF THE IMAGE STREAM SAYS NO?

Sometimes your Image Stream will refuse to answer a question. Such refusal can take the form of a character in the imagery shaking his or her head, or even some frightful, nightmarish image that seems to be warning you off.

Some of my colleagues believe that we should always respect such refusals by abandoning our query on the spot. I disagree. My experience has shown that refusals often do not originate in the unconscious but are instead manifestations of conscious fears or even laziness.

Sometimes a refusal protects some cherished but questionable belief. Any competent mental health professional will tell you that it is better to go ahead and test your beliefs than to coddle and shield them from possible debunking. Any belief worthy of preserving should be well able to defend itself in the free marketplace of ideas.

When you are confronted by a refusal, put the following questions to your Image Stream:

- "Who or what is saying no?"
- "What can I do right now to make myself ready or worthy to receive the information I'm seeking, without harming myself or others?"
- "If the no was caused by the form of my question, what form should I use in order to receive an answer?"
- "What should I be asking instead, in this context?"

When No Is the Right Answer

There are instances when a no really does seem to emanate from the right brain, perhaps because the inquiry we are pursuing may lead to harm, either to ourselves or to others.

Some experiences have suggested to me that my own Image Stream, at least, is reluctant to answer questions about how to make a better weapon or how to inflict defeat upon an enemy. My unconscious also appears hesitant to dispense advice that would help a person gain from someone else's loss, as in winning the lottery, picking the right horse, or even playing the stock market. Friends, colleagues, and acquaintances have had similar experiences, suggesting to me that our more subtle faculties are sensitive not only to our personal needs but also to those of humanity in general. Unless you can find a way to reframe such questions in a way that fully respects the needs of others, I recommend that you respect the "no" that follows.

CLARIFICATION QUESTIONING

Suppose you've just tried Thresholding a particularly difficult Image Stream. Perhaps you've run through the drill two or three times, each time getting new images but seeming to draw no closer to your goal of illumination.

It's time to try Clarification Questioning. Here's how it's done:

1. Identify a new Thresholding device, as described in Chapter 5.
2. Project a warm feeling of thanks to your right brain for showing you as much as it has and taking you this far. Then ask its help in clarifying your understanding of what it has shown you.
3. Reach out and touch your new Thresholding device.
4. While touching the device, say silently but "loudly" in your mind, "Show me the *same* best answer to this *same* question, but show it to me in an entirely *different* way."
5. Strengthen your contact with the Thresholding device by describing it in rich, multisensory terms.
6. Abruptly unveil the Answer Space, in whatever manner is appropriate to the Thresholding device you have chosen.
7. Step into the Answer Space and describe everything you see in rich, multisensory detail.

8. Use *inductive inference*. As you look around the Answer Space, ask yourself, "What's the same when everything is different?" No matter how different your surroundings may seem from your last Thresholding attempt, some crucial element will remain the same. Your answer lies within that unchanging element.

FEATURE QUESTIONING

You can put specific questions to virtually any object or person that appears in your Image Stream. Suppose you've just used inductive inference to locate an unchanging feature. Addressing the feature much as you would a Thresholding device, ask it any of the following questions:

- "Why are you in this scene?"
- "What is the meaning symbolized by your being in this particular position, relative to those (trees, bushes, people, or any other appropriate objects)?"
- "Why are you green (or whatever color or condition)?"
- "What is your role in this scene?"
- "What are you supposed to show me?"

In most cases, the feature will answer by shifting the scene, offering a new set of images for your consideration. Whatever image you see after asking your question, seize upon it, and begin describing it. In some cases, however, the feature may actually respond in words.

SEEKING WORD ANSWERS

Some people assume that a verbal answer will be clearer and less ambiguous than a visual one, but such is seldom the case. As with visual answers, the best verbal responses tend to be those that are least expected and most surprising. Such answers can also seem, at first glance, to be the most enigmatic. It's really no easier to interpret the words "lacquered box" than an image of a lacquered box.

Nonetheless, some people may feel more comfortable with words. There's nothing wrong with asking a feature to respond in words. It may or may not comply. Keep in mind that verbal answers tend to shift your perception back toward the analytical left brain and away from the pattern-seeking right brain that mainly guides your Image Stream. This shift could tend to break or weaken the spell of right-brain thinking.

Certain trainers, guides, and gurus whom I have encountered make little distinction between words and pictures when they are leading their clients or disciples toward "higher consciousness" through guided imagery. The imager might imagine himself being presented with the Book of All Knowledge or some such thing.

"Well then," says the trainer, "what does this book say?"

(Silence)

"Well, then," the trainer continues, "open to a page. What does the page *say?*"

(Silence)

"Well, what's the first *word?* Can you make out the first *letter?*"

And so on. Such coaching, in my opinion, can only succeed in wrenching the hapless imager away from the fruitful flow of right-brain imagery and back into a left-brain, verbal state of mind. Any results achieved at this point will lack validity, and many imagers will even be tempted to fudge in order to make their guru happy.

Turn Words to Pictures

The language of the right brain is the language of image, symbol, and metaphor. In general, even if it is your tendency to receive messages in verbal form, you should try, after reporting the words, to make up and describe pictures or scenes that would go with those words.

For example, if the word *freedom* pops in from somewhere, go ahead and report it, but then describe, say, the scene of an eagle floating high over a cliff against white swollen thunderheads in a summer evening sky—or whatever picture best evokes freedom for you.

Know When to Use Words

It should be noted that some people get very good information entirely through verbal answers. That's okay. If you keep getting word answers, run some self-tests. Alternate back and forth between verbal and visual answers and see which type provides greater ingenuity and insight. Go with whichever type gives you the best results. Recheck every month or so to see if the same method is still yielding better insights.

Use Pictures First, Words Later

There is, of course, no reason why you can't use both words and pictures in the same Image Stream. The general rule is pictures first, words later. While you are performing a Thresholding technique or questioning a feature in your Image Stream, wait for your answer to appear first as an image. Only after that image has fully congealed should you then listen intently for a word answer.

Once a picture has formed, it is harder for your conscious expectations to contaminate whatever words might follow. Words are most useful in clarifying a point that has already been made by an image.

Beware of Writing

Lucid dreamers report strange experiences when they try to read and write while dreaming. German physician Harald von Moers-Messmer, who experimented with lucid dreaming back in the 1930s, reported that letters in his dreams turned into hieroglyphs whenever he tried to focus on them.[3] Dr. Stephen LaBerge, the modern-day popularizer of lucid dreaming, finds that written words change form every time he looks at them. Ordinary dreamers do not report this difficulty quite so consistently. It seems peculiar to lucid dreaming, in which the words arise from the conscious will of the dreamer.[4]

I believe that lucid dreamers, like Image Streamers, are simply too much under the spell of right-brain thinking to make coherent sense of the written word. The two functions seem to have a mutually exclusive effect: When one connects, the other disengages. For this reason, verbal answers should be sought in the form of speech, not writing. I believe the spoken word encourages a more favorable balance between lateral brain functions.

THE JOKING ANALYST TECHNIQUE

"Many a truth is said in jest," goes the old saw. Humor is seated in the same right cerebral cortex that houses most of the higher and more sensitive faculties from which the Image Stream flows. Sometimes, engaging humor will provide the one extra connection we need for everything to link up and make sense in a particular Image Stream. A playful attitude helps us drop our guard and forget about being "right." It opens us to a wider set of possibilities.

When your spontaneous Image Stream ends without coming to any apparent conclusion, one device that may help you fish out meaning is the Joking Analyst. The technique proceeds as follows:

1. Make believe that you are Sigmund Freud, Carl Jung, Milton Erickson, or some other profound student of human symbology. Pretend that the Image Stream you've just reported is a dream that one of your patients has had.

2. Speculate freely on possible meanings of your "client's" dream, either on paper or on tape.
3. While talking, try to mimic the jargon and manner of the famous analyst you are impersonating, in the most humorous way possible.
4. Assume that every aspect of the dream is charged with meaning and pregnant with metaphor, even to a grossly exaggerated extent.
5. Have fun. Remember that it is much more important in this exercise that you be funny than that you try to be right.
6. Speak as rapidly as possible in order to short-circuit your internal Editor.
7. Sustain the flow for several minutes.
8. Go back over the tape or notes you made as the Joking Analyst. If you succeeded in letting go and getting into the role, your monologue will no doubt contain some key insights for interpreting your Image Stream.

THE CLARIFICATION REFLEX

As you become more fluent in the language of the right brain, your Image Streaming experience will evolve more and more toward an active dialogue between your conscious and unconscious minds. You will resort reflexively to Thresholding, Feature Questioning, and Clarification Questioning in response to almost any puzzling development during the Image Stream, and you will execute them quickly, on the spot, without breaking your stride.

To develop your Clarification Reflex, get in the habit of asking the following questions whenever your Image Stream seems to grow confusing or obscure:

- "Please show me, in a new picture, whatever I haven't yet fully understood from your previous picture."
- "What further question should I ask now?"

BE A GIDEON

A well-known Bible story relates that an angel of God once visited a man named Gideon, calling upon him to take up the sword to free Israel from the yoke of the Midianites. Gideon found it hard to believe that God would choose him for such a task. Instead of blindly obeying these instructions, Gideon put his apparent perception of God to the test, respectfully challenging Him not once but twice to confirm His instructions with a miraculous sign.

Instead of punishing Gideon for his impudence, God rewarded him first with apparent miracles involving a fleece, then with a miraculous victory. Gideon defeated the Midianite army with only 300 men, established a peace that lasted forty years, and was called a "mighty warrior." Doubting Thomas, who in a later book of the Bible demanded proof of Jesus' resurrection, was rewarded by the Church with sainthood.

The Bible is often misconstrued as a call to blind obedience. In fact, I believe that the stories of Gideon and Thomas clearly encourage the faithful to test and question what appears to be the Word of God. In so doing, we are questioning not God but the accuracy of our own perception.

In the same spirit, I urge you to question every message from your Image Stream. I can think of nothing more dangerous or frightful than the assumption that Image Streams are infallible. The unconscious remains a mysterious and unknown force. We are far from understanding even the tiniest portion of its workings or motivation. At best, the messages from the Image Stream should be looked upon as food for thought, a reservoir of ideas that are only as reliable as our ability to test them in action.

My rule of thumb is, the clearer the message seems, the more urgent your need to "be a Gideon." One way to accomplish this is to ask your Image Stream, "How can I make sure that what you are telling me is so?" or "How can I determine whether this is true?"

Ultimately, though, you will have to test your Image Streaming messages exactly as you would test any other thought or inspiration that came into your head—through reason and action. Be an active skeptic. Rely on your intuition. Never hold to any course of action that seems wrong or doubtful just because your Image Stream "told you so." Use your own good judgment.

A NOBLE ENTERPRISE

Since time immemorial, the ability to interpret dreams and visions has been cherished by kings and emperors. Fluency in the language of the right brain remains, to this day, a rare and precious gift, imparting to its wielder extraordinary power. Yet the regimen laid out in this chapter can lead you rapidly and naturally to a high degree of such fluency. I urge every reader to persist in these exercises. Interpretation may well prove a more difficult skill than many others offered in this book, yet no other discipline will offer rewards in such abundance. The incomparable advantage of finding a portable, ready-made, genius-level problem solver installed conveniently between your ears is surely worth your effort and attention.

CHAPTER 7

THE POWER OF QUESTIONS

Einstein once remarked that if he were about to be killed and had only 1 hour to figure out how to save his life, he would devote the first 55 minutes of that hour to searching for the *right question*. Once he had that question, Einstein said, finding the answer would take only about 5 minutes.

Until now, we have looked only at methods for *answering* questions—less than 9 percent of the work, according to Einstein. What about the other 91 percent? How can we use Image Streaming to lead us to the right questions?

One way, which we recommended in the last chapter, is to simply ask your Image Stream, "What is the best question for me to ask at this point?" Your Image Stream will answer this question readily. But until you have become fluent in the language of the right brain you may have to go through several rounds of probing and clarification before the question becomes plain to you.

There is a simpler way, which we are about to reveal, that experience has shown to be quick, easy, and highly effective. It cuts through all conscious expectations, answering only your most pressing and urgent questions rather than those questions that you *think* are most pressing. The answers you receive through this method will be the purest and most spontaneous of all. Your left brain will find it physically impossible to intrude with its notions of what the answer *ought* to be.

In order to use this incredible technique, however, you must take a conceptual leap somewhat larger than those you have already been asked to take.

HIDDEN QUESTIONS

The method to which I am referring is the use of Hidden Questions. Write out six questions on separate slips of paper, shuffle the papers, and then select one slip at random. *Without looking at the slip*, consult your Image Stream. It is impossible for you to know consciously which question you have selected, but your Image Stream knows. In most cases, the answer you get will prove to be an uncanny fit to the chosen question.

Using Hidden Questions is the best way I know of to bypass the Editor. If your left brain doesn't even know the question, you can't possibly anticipate or plan a conscious answer. But how does it work? How can your Image Stream read a question that is concealed from your conscious mind?

The Butterfly Incident

At my seminars, I always strive to encourage a stimulating, open-ended atmosphere. Sometimes participants plunge spontaneously into discussions that range far afield from the subject of Image Streaming.

This happened at a seminar I conducted in Ravenna, Ohio, in 1981. Somehow, we all got talking about the eternal question of what happens to people after death. The hard-core atheists in the group insisted that everything ends with physical death. Others argued for eternal life, citing traditions that ran the gamut of just about every known spiritual and religious faith. As the debate grew more heated, I tactfully suggested that we call time out and learn some new Image Streaming techniques. Soon, to my relief, I had gotten the group engrossed in a Hidden Question exercise.

Without telling anyone what I was up to, I slipped a trick question into the pool of concealed queries: "Does consciousness continue past the point of physical death?" Only moments before, participants had been locked in conflict over that question. But when the same query turned up during Hidden Questions, a remarkable consensus emerged. Remember that no one knew the concealed question until the exercise was over. Nevertheless, while Image Streaming, every one of the thirty-four participants had received some form of the same answer. Each had had a vision of a million butterflies rising from a meadow into the sun!

ACTIVE SKEPTICISM

Readers will react to this story in different ways depending on their beliefs. Some will declare it to be a hoax, a lie, or at best a coincidence. Others will immediately assume that some psychic or telepathic force was at work in the group. Both assumptions, in my opinion, jump the gun. *Active skepticism* is the best response to an unknown phenomenon.

When Wilhelm Roentgen announced in 1895 that he had discovered a mysterious new form of energy that could pass right through people's bodies and make photos of their bones, some scientists were understandably skeptical. But Lord William Thomson Kelvin, President of England's Royal Society, went beyond mere caution. He declared with perfect confidence that "X-rays will prove to be a hoax."[1]

Around the same time period, a member of the French Academy of Sciences made this report to his learned colleagues: "Gentlemen, I have personally examined Mr.

[Thomas] Edison's phonograph and I find that it is nothing but a clever use of ventriloquism."[2]

Kelvin and his French colleague were *absolute skeptics*. They acted as if anything outside their current worldview was, by definition, impossible. Despite its pretensions to objectivity, absolute skepticism opposes scientific advancement. Virtually every genius in history has had to fight hordes of sneering absolute skeptics in order to make any discovery. In contrast, the *active skeptic* aggressively seeks proof of unexplained phenomena while maintaining an open mind. This is the attitude of the true scientist. I encourage all my readers to be active skeptics as they proceed through this chapter.

TELEPATHY?

Some years ago, a group of parapsychologists thought that they had uncovered conclusive proof of telepathy. They had been testing the ability of experimental subjects to "guess" which concealed card a researcher was holding. Certain test subjects gave so many accurate responses that mere guesswork was ruled statistically impossible. Those subjects must be using telepathy, the parapsychologists reasoned.

Fortunately, there were some active skeptics on hand to double-check the data. It turned out that the "telepathic" subjects were not reading the researchers' minds at all. They were reading body language. Somehow, they had figured out a way to tell which card had been selected just by the way the researcher *looked* at it.[3]

The Sensitive Mind

Enthusiasts of telepathy were keenly disappointed by these results. Nevertheless, the experiment revealed a degree of perceptiveness in the test subjects that was hardly less surprising than the ability to read minds.

For thousands of years, Polynesian sailors have known how to navigate by feel alone. You can take a traditionally trained outrigger pilot, blindfold him, and place him in the water hundreds of miles from home. Just by feeling the ocean currents on his body, he will be able to pinpoint his location with great accuracy.[4]

Such discernment is not in the least bit psychic or paranormal, yet it seems hardly less marvelous than clairvoyance or telepathy. Like the crafty test subjects in the aforementioned ESP experiment, Polynesian sailors have managed to harness perceptions so subtle that most of us aren't even aware that we have them. The Butterfly Incident at my 1981 seminar may have resulted from just such enhanced attentiveness. Perhaps those thirty-four participants unconsciously tuned in to faint, subaudible cues and patterns of behavior throughout the room. Pending further research, we can only guess.

One thing is sure. Whatever the cause of these supersubtle perceptions, they are extraordinarily powerful. If we can learn to direct them to the solution of specific problems, we will have taken a great leap toward attaining true genius.

ALL AT ONCE

Like many geniuses, Wolfgang Amadeus Mozart claimed that he wrote his musical compositions entirely in his head, crafting every chord and arpeggio to perfection before setting a single note to paper. Mozart would amaze his contemporaries by performing such feats as "writing" music between shots of a billiards game or scribbling down the overture to the opera *Don Giovanni* mere hours before its opening performance. On such occasions, Mozart explained, he was not writing the music at all but simply taking dictation from the already finished piece in his head.[5]

In a letter written in 1789, Mozart explained that, before committing any work to paper, he first surveyed the whole in his mind, "like a fine picture or a beautiful statue." Mozart did not review his creations as an orchestra might play them, one measure after another. Instead, he took in the whole "at a glance." "Nor do I hear in my imagination the parts successively," he wrote, "but I hear them, as it were, all at once. What a delight this is I cannot tell!"[6]

Ingenious Thought

Mozart's creative method clearly transcended the ordinary. How can one "survey" an entire symphony "at a glance"? How can one hear it "all at once" rather than "successively" over time? This is a riddle as perplexing as five-dimensional geometry. Yet this bizarre mode of thought came as naturally to Mozart as the piano keys to his fingers.

Truly ingenious thought occurs in a realm wholly unlike that which our conscious minds can grasp. Great geniuses routinely disengage from rules of ordinary perception that most of us think unbreakable, playing havoc with commonplace notions of time, space, and form.

PURE IDEAS

In his masterpiece, *The Republic*, written in the fifth century B.C., the Greek philosopher Plato speculated that there were really two worlds: the material world that we apprehend with our senses, and the world of pure *ideai*—ideas. Only in the realm of ideas do we see things in their true form, said Plato. Music, poetry, painting, and mathematics are but crude attempts to capture the transcendent beauty and order of the ideal realm.[7]

Plato's Cave

Plato likened our everyday world to a dismal cave in which people were chained by their necks, unable to turn their heads or to see anything except the cave wall directly before them. There, on that wall, they beheld a play of shadows cast by other people deeper in the cave who moved about in the firelight, out of sight. In their ignorance, the chained prisoners believed these shadows to be the only "real" things in the world. Yet if they could only turn their heads, they would see the fire and the people who formed the shadows. If they could throw off their chains, the prisoners could even make their way outside and feast their eyes upon the glorious, sun-filled world.

All of us are chained in such a cave, said Plato, seeing only shadows upon the wall and imagining those shadows to comprise our whole universe. Yet ". . . the soul of every man does possess the power of learning the truth," Plato wrote, "and the organ to see it with." If we could but free our perceptions from their shackles, Plato taught, we could at last "contemplate reality and that supreme splendor which we have called the Good."[8]

Ecstasy

During the Renaissance and Enlightenment periods, philosophers revisited Plato's cave. The Neo-Platonists of the Renaissance believed that great artists gained a glimpse of the ideal realm during the throes of creation. Jean Jacques Rousseau speculated that any mortal who made contact with Plato's ideal realm would be consumed by *ecstasy.* Perhaps it was this ecstasy to which Mozart referred when he declared, "What a delight this is I cannot tell!"[9]

Plato Vindicated

Until very recently, speculations such as those of Plato and Rousseau were consigned to the realm of mysticism. There was no realistic way of proving them. But things have changed in the twentieth century. The advent of quantum physics has brought forth powerful new tools that may yet grant us our first scientifically measurable glimpse outside the cave. In the process, it may also have laid bare the mysterious engine of creation at work in the mind of a Mozart.

SPOOKY ACTION

In the 1920s, physicists discovered that subatomic quanta, such as photons and electrons, have a protean ability to shift their form in an instant. Measured one way, they behave as particles. Measured another way, they act like waves. Even more confusing, if you measure the velocity of a quantum particle you cannot measure its mass, while if you measure its mass you cannot measure its velocity. This confusion led Danish physicist Niels Bohr to speculate that quanta did not really possess such properties as mass, velocity, and wave state but only acquired these characteristics temporarily, in response to physicists' efforts to measure them. In their natural state, said Bohr, quanta existed only in a formless void of *potential* characteristics.

Albert Einstein could not accept this explanation. What if a particle composed of two protons suddenly flew apart, Einstein asked, sending its two component protons hurtling into space? The law of Conservation of Momentum states that if we know the momentum of one we can calculate the momentum of the other. But according to Bohr's theory, neither proton has any momentum at all until the instant we attempt to measure it.

Bohr's theory suggested that if we measured the momentum of one proton (say, proton A), our very act of measuring proton A would simultaneously "cause" proton B to acquire a corresponding momentum—even if the two protons had flown, in the meantime, to opposite ends of the universe! To accomplish this, the two protons would have to communicate or coordinate their respective momenta across the universe at faster-than-light speed, almost as if they were telepathically connected. Einstein scoffed at such "spooky action at a distance." Bohr and his followers must have overlooked some "hidden variable," he insisted in a 1935 paper cowritten with Boris Podolsky and Nathan Rosen. Bohr's theory was simply incomplete, Einstein concluded.[10]

THE AQUARIUM MODEL

One of Einstein's disciples, David Bohm, proposed a solution to this paradox. He agreed with Einstein that it would be both "spooky" and unlikely for two protons to communicate instantaneously across the universe. Instead, Bohm proposed that quanta and all their peculiar actions were but reflections of a deeper order underlying the visible universe. He likened quantum particles to a fish in an aquarium. Suppose we viewed the fish through two TV monitors hooked up to two video cameras, each pointed at the fish from a different angle. Through our monitors, we could see two different images of what appeared to be two different fish. When one fish turned, the other fish turned simultaneously, as if the two were mysteriously connected. Only by looking directly at the aquarium could we learn that the two images were really of the same fish, viewed from different angles (see Figure 7.1).

Bohm's aquarium, like Plato's cave, was a metaphor for the limitations of human perception.[11] In normal life, we see only the perplexing images on the video monitors. But suppose we could look directly at the "aquarium." Suppose we could unchain ourselves and emerge from Plato's cave. Unhobbled by our five feeble senses, what would we really see out there?

THE IMPLICATE ORDER

How would your favorite computer game look if you had no PC to translate the software into pictures? How would a phone call sound if you had no receiver? The words and

Figure 7.1 Physicist David Bohm likened the world beyond our senses to an aquarium hidden from our sight. If we view a fish through two different video cameras, it appears to be two different fish that mysteriously move in synchrony with each other, much as quantum particles seemingly interact with one another across improbably vast distances of space and time. Einstein called such quantum behavior "spooky action at a distance." Bohm suggested that such "spooky" interactions may actually involve a single quantum particle, which—like the fish in the aquarium—only seems to be two separate particles in two different places due to our imperfect grasp of the "implicate order" (Bohm's phrase for the unseen world).

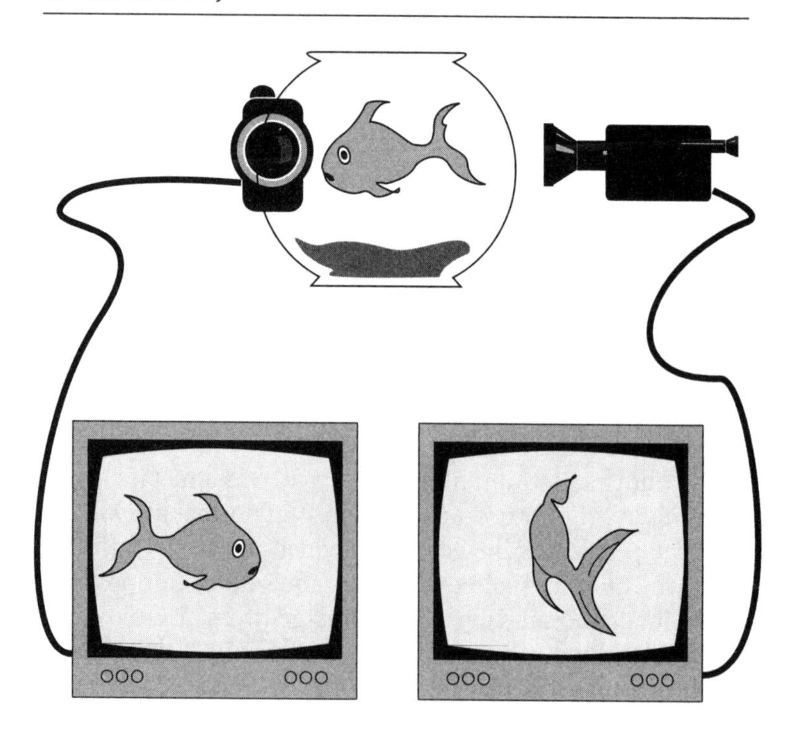

pictures would dissolve into waves of invisible energy. Indeed, we could not perceive them at all with our ordinary senses. Like the electromagnetic impulses of modern telecommunications, our material world takes on coherent form only when it makes contact with an appropriate receptor, such as one of our five senses.

At least that's what David Bohm thought. He hypothesized an implicate order to the universe, in which all that we see is encoded in patterns of pure energy. Only the peculiarities of human perception translate this mass of writhing energy into the familiar explicate forms of our three-dimensional universe.

The Holographic Principle

If you drop a pebble in a pond, ripples will spread out from the pebble in concentric circles. Drop three pebbles, and the circular waves they create will crisscross, making an *interference pattern* on the water.

Now imagine that you could freeze that pond instantly, capturing the interference patterns in an instant of time. If you broke off part of the frozen pond, the interference patterns encoded on that chunk of ice would give you all the information you needed to calculate exactly where the three pebbles had fallen.[12] (See Figure 7.2.)

In 1947, scientist Denis Gabor discovered a way to encode three-dimensional objects on photographic film, much as we encoded the dropping of the three pebbles. Gabor's process, which also used interference patterns, came to be called holography.[13]

To make a hologram, you fire one laser beam—called the "data beam"—directly at an object so that it reflects onto a photographic plate. At the same time, you fire a reference beam directly at the plate. When the two beams converge, they create interference patterns, which are then captured on the photographic plate.[14]

If you look at the plate, you see not a picture but a complex pattern of wavy lines—the interference pattern. To get a picture, you have to shine a laser beam back onto the plate. Its reflection forms an exact three-dimensional image of the object, hovering in space at the same position relative

Figure 7.2 The smallest piece of a hologram contains sufficient information—in the form of "interference patterns"—to project the whole image. A similar effect can be attained from frozen ripples in a pond.

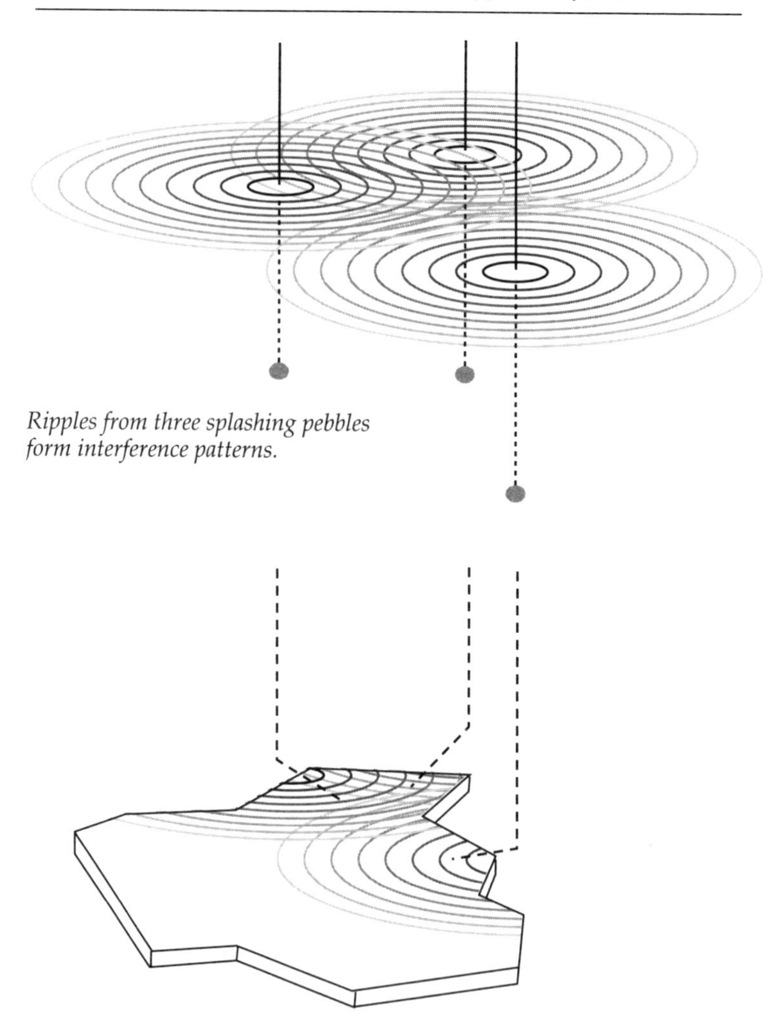

Ripples from three splashing pebbles form interference patterns.

Freeze the ripples and break off any piece. The patterns on that piece will enable you to calculate where each pebble fell in the pond.

to the plate as was the original object when the hologram was taken.

Like the chunk of ice, any piece of the photographic plate that is broken off will contain all the information necessary to enable the laser beam to reconstruct the entire three-dimensional hologram. If you break the plate into ten pieces, you will have ten separate holograms.[15]

Nonlocality

Bohm suggested that the familiar world around us is similar to a holographic projection, reflected from the implicate order much as a hologram is reflected from a photographic plate. Just as the hologram only *seems* to hover in space at a particular location, so the objects in our universe only seem to occupy certain specific localities. Because the totality of space is completely frozen or encoded in the implicate order, every location in the universe is equivalent to every other location. The very idea that space is separated by inches, miles, light-years, and parsecs is but an illusion, as ephemeral as the shadows in Plato's cave.

As we see from the odd behavior of subatomic particles, the most unreachable distances of space can indeed be traversed in an instant through slight adjustments in the implicate order—analogous to a slight change in the orientation of the goldfish in Bohm's aquarium. This is the principle of nonlocality: Distance, no matter how vast it may seem to us, really doesn't matter.[16]

The Illusion of Time

Time is another holographic illusion, according to Bohm. Our senses tell us that time proceeds in linear progression, from the past to the present to the future. But in the implicate order, all that has been or ever will be is fully encoded right now in the superhologram.

Bohm speculated that our enjoyment of beautiful music gives us a glimpse at the implicate nature of time. Although we hear music in chronological succession, with one note or chord following another, we do not *appreciate* music in such linear order. Instead, each successive note causes what Bohm called an "active transformation" in our perception of all the notes that came before. It likewise creates a strong anticipation of all the notes that come afterward. We appreciate the piece only in its entirety, with beginning, middle, and end all interacting at once in our minds.[17]

When Mozart reviewed his finished pieces "all at once," he may have employed perceptions far truer than our own. Bohm's view would suggest that at least part of Mozart's genius rose from his ability to apprehend music at the deeper, implicate level of reality.

THE PSI QUESTION

Bohm's holographic model of the universe suggests a plausible framework by which we might understand such claimed human faculties as clairvoyance and telepathy—which parapsychologists call psi phenomena. If time and space are indeed connected at the implicate level, then it follows that human perception may sometimes transcend ordinary bounds of space and time. Interested readers can find a marvelous elucidation of these ideas in Michael Talbot's books *The Holographic Universe* and *Beyond the Quantum*.

Today, the Psi Question is no longer relegated to the academic fringe. New evidence has forced modern science to take it seriously. In 1969, the Parapsychological Association was finally admitted to the American Association for the Advancement of Science (AAAS) marking parapsychology's acceptance as a legitimate science.[18] In 1982, Alain Aspect, Jean Dalibard, and Gérard Roger at the Institute of Theoretical and Applied Optics in Paris succeeded in producing "spooky action at a distance" between two photons in the laboratory, thus empirically supporting the theory that nonlocality is a real phenomenon.[19,20,21] Five years later, in 1987, physicist Robert G. Jahn and psychologist Brenda J. Dunne of Princeton University's Engineering Anomalies Research Laboratory published compelling evidence that test subjects had been able, through the power of focused thought, to influence the output of random-number generators—strong evidence of a "mind-over-matter" effect.[22]

SUPERSTRING THEORY

Physics has evolved far beyond the arguments over quantum theory that dominated the 1920s and 1930s. One new branch of speculation, called Superstring Theory, may provide us with an explanation for paranormal phenomena as compelling as Bohm's aquarium model. It asserts that the universe actually exists in ten dimensions. Our familiar, four-dimensional universe (three dimensions of space and one of time) presumably came into being trillions of years ago, when a single point of infinite density exploded in what is known as the Big Bang. But for some reason, say the Superstring theorists, the other six spatial dimensions failed to explode outward and remained tightly packed in the center of the universe.

If you were suddenly to come ajar or "rotate" yourself into one of these "underdeveloped" axes, you would find yourself in the odd position of being a giant, infinitely larger than the universe. You would then have intimate contact with every point in the universe at once, and every tiny movement you made would literally have cosmic repercussions. From such a god-like perspective, you would no doubt witness some very strange phenomena indeed. Perhaps it is through such chance rotations that some people gain fleeting glimpses into realities different from our own.

ALTERNATIVE EXPLANATIONS FOR THE HIDDEN QUESTION EFFECT

This book began with the story of Bob S., the young man whose Image Stream warned him that his fiancée was in danger. Until now, we have assumed that Bob S. must, at some point, have physically glimpsed the damage to his fiancée's car tire and registered the information subliminally. This is the safest and most conservative interpretation, yet it is by no means the only one.

Can it be that, by short-circuiting our analytical faculties, Image Streaming opens the mind to more direct apprehension of the implicate order? It's hard to say, but the possibility is intriguing. In the fifteen years that I have been teaching Image Streaming in my seminars, I have seen enough to conclude that there is far more to the process than meets the eye. Nevertheless, we should consider the many alternative explanations for some of Image Streaming's more remarkable results.

It is quite likely, for example, that the unconscious can easily unravel the hand movements involved in shuffling the slips of paper we use for Hidden Question. Anyone watching the shuffling process might thus have a subliminal awareness of exactly which question has been selected. Those who didn't watch could still pick up subtle cues from the facial expressions and body language of other participants, as the pseudotelepaths did in the ESP experiment.

Hidden Question doubtless draws much of its uncanny power from the phenomenon of "force-fitting." Edward DeBono, top corporate trainer, creativity consultant, and author of *Lateral Thinking*, has pointed out that many problems can be solved through what he calls the "provocative operation" or P.O. DeBono suggests that if you approach a problem from an unpredictable angle—*any* angle—you will come up with a creative solution. One of DeBono's provocative operations is to open a dictionary, select any word at random, then brainstorm all the possible relationships between that word and the chosen problem.

DeBono has found such random problem solving, which other creativity experts call force-fitting, to be highly effective. Indeed, Leonardo da Vinci's technique of visual free association could be seen as a form of force-fitting.

Such explanations demystify some, but not all, of Image Streaming's "spookier" results. Fortunately, we don't need to explain them in order to use them, any more than Mozart needed to explain his peculiar manner of composing music.

For now, let us play the active skeptic and explore the many uses of Hidden Question. Don't let its puzzling nature put you off. Where would we be today if no one dared flick on a light switch until after fully grasping electrical theory? Most of us would be stuck, quite literally, in the dark. Whatever explanation for Hidden Question we choose to believe, we can not deny the power of the technique. It is there for us to use, and we should take advantage of it.

THE HIGH THINKTANK EFFECT

The power of Hidden Question lies in its efficiency. In other Image Streaming methods, much of our time and effort is spent trying to squelch or fight the inner Editor. Hidden Question bypasses the Editor entirely by presenting questions directly to the right brain, giving the left brain no chance whatever to contaminate the data.

For this reason, I have found Hidden Question to be peculiarly suited to brainstorming the Big Questions of politics, metaphysics, philosophy, and social vision in large or small groups. More than any method I know, it seems to draw out consensus around issues that normally divide people very sharply. I call this the High Thinktank Effect.

High Thinktank also cuts to the quick in dealing with tough, interpersonal issues such as those that arise in a marriage or family, where years of ingrained habits have left people with the mistaken notion that all possible alternatives have been tried and all positions staked out in advance.

THE HIDDEN QUESTION PROCEDURE

So far, Project Renaissance has developed at least fifty different techniques for attaining the High Thinktank Effect. All are derived from the basic Hidden Question procedure, which follows:

1. Think up at least six different questions. Six questions will overwhelm your Editor and cause it to give up trying to guess which question you are answering. With fewer than six, The Editor will still have a good shot at guessing and may try to butt in and tell you what your answer should be.
2. Make sure your questions are all widely different from one another, so that no two questions can be answered in the same way. Some might be highly personal or practical; others might revolve around work, friends and family, or community, national, and world issues. Some might even be profound philosophical questions. Each time one of these questions is answered during your Hidden Question session, you should replace it in the question pool with another question in the same general subject category. This will ensure that you keep sufficient variety in your pool to outsmart The Editor.
3. Write each question on a separate slip of paper or index card. Fold the papers in half or turn the index cards over so the blank side faces up.
4. Swirl the papers or cards around randomly on the table, blank sides showing. Continue shuffling until your conscious Editor cannot possibly guess which is which. No matter how much you shuffle, your more sensitive faculties will still know exactly which is which.
5. Pick out one of the slips.
6. Without peeking, hold the slip in your hand or even press it against your forehead.
7. Close your eyes and generate three quick images or picture answers in succession. After receiving each picture answer, project thanks to your right brain and ask it to help you by showing you that same answer in a differ-

ent way. Each image gives the same answer to that same question, but *shows* it in an entirely different *way*.

8. Take no more than 15 to 30 seconds to record each picture answer on tape or paper or to a partner. You might find it convenient to jot down or even sketch each picture answer on a note pad.
9. Do not peek at any question until the whole process is complete. Otherwise, your Editor, by a process of elimination, might figure out which questions are left in the pool.
10. Determine which elements seem the same or similar among the three picture answers for the question you're working with. Similarities can be very subtle. You might notice the color green in each image, for example, or the absence of color, or triangular shapes. There might be a certain type of motion or lack of motion, people or no people, or a certain emotional impression. Write down your observations.
11. Finally, look at your question. This will often present a clear "Aha!" experience when you note what the three different picture answers have in common.
12. Where you do not get a clear "Aha!," you can try a bit of force-fitting. Play Joking Analyst and brainstorm all the possible ways that answer set could in fact answer the particular question.
13. If the meaning still eludes you, choose one of the picture answers that seems most provocative to you and use it as a Thresholding device. We recommend doing only one question in a session this way. Replace the finished question with another, for next time—and keep your question set where you can often turn to it.

THE QUESTION SANDWICH TECHNIQUE

Another excellent way to get much understanding by High Thinktank methods is this next procedure—Question Sandwich.

1. Buy a supply of small envelopes.
2. Prepare your sandwich questions. On separate scraps of paper or on index cards, write out single questions and put each question into a different envelope. You need at least six of these to "randomly" choose from. It is best for you to make up your own questions. But if you're stuck, you can choose from the following:

   - "For what opportunity should I be most alert today?"
   - "What is the most valuable action I can take today?"
   - "What is the most important new idea, perception, or observation I should grasp today?"
   - "In what way can I best serve others today?"
   - "How best can I create more wealth today?"
   - "For what problem should I be most on the lookout today, and what is my best response to it?"
   - "How can I best help______(my spouse, my friend, my child, my employee, whomever)?"
   - "What original discovery can I begin making today?"
   - "What is the best question I should ask now, and what is its best answer?"
   - "How best can I advance my goals today?"
   - "What is the one most important thing I should accomplish today, and how best can I achieve it?"
   - "What is the most important thing for me to unlearn today?"

   You can no doubt think of other—and perhaps better—questions in the same vein. All should be time-sensitive, in that they will tend to have different answers on different days. You may, of course, insert one or two highly specific questions in the stack as well.
3. Once all questions have been inserted in their envelopes, shuffle the deck thoroughly.
4. Place the stack by your bedside before going to sleep at night.
5. In the morning, pick an envelope. Without opening the envelope, obtain three picture answers from your Image Stream.

6. Record these answers on the outside of the envelope, on a notepad, or on a tape recorder.
7. Compare your three images to determine what they have in common.
8. Open the envelope and read the question. If the relationship between image and question is not immediately apparent, try force-fitting all possible relationships.
9. If the "Aha!" connection still eludes you, let it rest for a while. As you go through your day, the "Aha!" experience may jump out at you unexpectedly.
10. When you've finished with the question, remove it from the old envelope, seal it in a fresh one, return it to the deck, and reshuffle. You may want to collect your used envelopes in a file to keep a record of how many "hits" you are getting from your sandwiches.

The Hybrid Technique

You might find it convenient to combine the Question Sandwich technique with your daily 10-minute quota of Image Streaming. Simply write out your six concealed questions, select one, and then freely Image Stream for at least 10 full minutes on that hidden question. At the end of your session, read the question and proceed with appropriate follow-up. You might want to use a timer, alarm clock, or partner so you don't have to keep track of the time and can thus relax more fully into your Image Stream.

Make It Fun

Don't let Question Sandwich become a grim daily chore. An overly serious approach will not only encourage you to abandon the exercise after a few days, it will also dampen the ability of your subtler faculties to achieve useful answers.

Treat Question Sandwich as a game. Take a lighthearted, half-joking approach, as if you are opening fortune cookies with your friends in a Chinese restaurant. One way to add to the fun is to use "hand dowsing." Run your hands over the envelopes and choose the one that makes your fingers tingle or gives you some other special feeling. Also important is rationing yourself to one question per session in order to keep the process from becoming a burden on your time.

You can add extra convenience to Question Sandwich by placing the set of questions in easily accessible places. If you are a regular coffee or tea drinker, set those questions near your coffee stand. Or put them by the chair you sit in to read your daily paper or next to the "In" box on your desk. You can even carry them in your pocket throughout the day. Whenever you have a free moment, simply place your hand on one of the cards at random and glance inward to your Image Stream to retrieve three alternate picture answers.

Easy access will encourage you to sandwich in questions several times during the day, until it becomes a well-established and systematic habit.

THE GAME OF ORACLE

When you engage in Hidden Question with other people, the group atmosphere brings out a lively social energy and a great deal of humor—both of which seem to enhance the ease and accuracy of the picture answers. Because of the festive mood they create, I call such group exercises the Game of Oracle.

1. Find a partner.
2. Each of you make up a number of questions, ranging in number from two to ten or more. They should be open-ended, not yes-or-no, questions. In other words, don't ask "Shall I ask my boss for a raise?" Instead ask "What's the best way to earn a raise from my boss?" Make each question very different from every other question—so different that you think there is no way your partner's right brain could predict that you would be asking that question.
3. Take turns silently "reading" your question to your partner. In groups of three or more, simply ask one person to "read" the question while the others answer in turn.

4. After each question, your partner should take a few seconds to generate three picture answers. If you have three or more partners, each partner can generate an image or two instead of one generating all three.
5. Take a few minutes to discuss the picture answers together and to determine the common elements in each.
6. Reveal the question to your partner. Compare your answers with the question, noting especially the common elements among them, to make an interpretation.

The Game of "Phnoracle"

Oracle works quite well over the phone. A special energy comes from in-person encounters, but telephoning is a lot quicker. If some important issue or question arises, you can phone a friend—preferably someone with whom you have had previous Oracle experience—and come up with answers on the spot. I call this the Game of Phnoracle.

Forget the First Commandment

We have already noted several exceptions to the first commandment of Image Streaming: Thou shalt not interpret for thy partner.

In the Game of Oracle, there is no first commandment at all. The picture answers you get in the Game of Oracle, unlike those in other Image Streaming situations, are as likely to be given in the questioner's code as in the code of the answerer. Interpretation, therefore, becomes fair game for *anyone*.

PORTABLE QUESTION BANK

In order to keep your question pool well-stocked, you might want to acquire the habit of collecting questions throughout each day. Carry a pocket notepad or a pack of index cards and routinely jot down each question or problem that comes to mind. Whenever you have collected a half dozen or so question cards, fold each one over and shuffle them all into your Hidden Question stack.

PUSHING THE ENVELOPE

Some readers may want to test their subtle faculties by experimenting with ESP questions—queries that cannot possibly be answered by the left brain. I am not talking here about mind-twisting Zen koans but rather perfectly logical, straightforward questions, the answers to which your conscious mind simply has no access. The results can be both amusing and astonishing.

A few suggestions follow:

- "What will be the outcome of (some pending enterprise)?"
- "What's the biggest surprise I'm going to have next week?"
- "What surprising discovery will be revealed by (some pending scientific experiment or expedition, such as an upcoming flyby of Neptune)?"
- "What would save so-and-so's life?"

FLEXING YOUR PSI MUSCLES

Because it can be easily blended into a busy day, Question Sandwich has become the most popular and effective of the Hidden Question techniques. Within a year after we introduced Question Sandwich into our Project Renaissance arsenal in 1987, it became evident that this technique was nearly as potent a Pole Bridger as straight Image Streaming.

The concealment of the questions forces your mind to work with extraordinarily subtle perceptions. It matters little whether you view these as psi perceptions or as acute natural sensitivities similar to those exhibited by Polynesian sailors. By cultivating them daily, you bridge special poles in your brain that are rarely exercised in other circumstances.

The study of such acute faculties is so new that we can only speculate as to what benefits may accrue from long practice. If the ESP enthusiasts had not thrown away in disgust their first ten years of work—just because they had gotten different results from what they had wanted and expected—we'd now have several more decades of research built up in this invaluable topic of subtle sensitivities.

As it is now, Question Sandwich and straight Image Streaming seem to offer complementary strengths. Each technique bridges some poles in your brain that the other does not. Used together, these methods will catapult you to levels of mental agility you never thought possible.

CHAPTER 8

MODEL THINKING

The late 1970s were an exciting time in the field of accelerative learning. For years, Communist governments had funded high-priority research programs studying methods of subconscious learning. Researchers such as Georgi Lozanov and Vladimir Raikov had achieved astonishing results, but little information seeped out through the Iron Curtain. Only after the publication of *Superlearning* by Sheila Ostrander and Lynn Schroeder in 1979 did these techniques become widely known in the West.[1] Until then, most researchers in the United States were thrashing around in the dark, trying to piece together from rumors, gossip, and incomplete reports the precise methodologies used by our colleagues in the Soviet Bloc. At that time, even American breakthroughs, achieved by pioneers such as Jean Houston, Robert Masters, and the great hypnosis researcher Milton Erickson, were poorly covered in the scientific literature and almost as inaccessible as the Soviet results.

It was during those early exploratory years that I first stumbled upon a technique that has become one of the most powerful tools in the arsenal of accelerated learning—

Model Thinking. This non-hypnotic technique enables a person to borrow the identity and talents of any chosen role model from history, such as a Leonardo da Vinci, a Michelangelo or an Einstein, thus unleashing stores of hidden talent from within.

THE ARLINGTON EXPERIMENT

In March 1977, a group of us decided to conduct an experiment in a friend's apartment in Arlington, Virginia. Our approach was hit-or-miss. We would try out a variety of the new enhanced learning methods, using ourselves as guinea pigs. Although no one at that time had published reliable accounts of the exact procedures, we reconstructed them as best we could from the few scraps of information available in odd corners of the scientific literature.

It was our hope that at least one person in our little group of 17 might gain some slight benefit from these reconstructed methods. Then we would zoom in on that individual, examine closely what had happened, and try to replicate the result until, at last, we might succeed in developing a working methodology. In hindsight, I don't think any of us really expected dramatic results.

We were completely surprised. Nearly every technique we tried produced striking results for almost everyone in the group. Only years later, through my work in Image Streaming, did I begin to grasp the theoretical mechanism behind these methods. But at that 1977 workshop the results spoke for themselves.

Perhaps the most memorable result was the experience of one participant I will call Mary. Like all of us, Mary had agreed to embark upon some new learning experience just prior to the workshop. She chose the violin. Mary had her first lesson one week before our experiment. Until that time, she had never touched a violin in her life.

The week following our workshop, Mary had her second violin lesson. Because she worked full-time as a secretary in a Washington, D.C., office, she had only a moderate amount of time to practice. Nevertheless, after Mary had played for a few minutes at her second lesson, her astonished instructor announced that he was going to enroll her in his advanced class! At our second experimental workshop, two weeks later, Mary gave us a very fine concert with her violin.

THE RAIKOV EFFECT

Mary owed her precocious ability to the Raikov Effect. Soviet psychiatrist Dr. Vladimir Raikov developed a method called Artificial Reincarnation in which he used deep hypnosis to make people think they had actually, physically *become* some great genius in history.

When he "reincarnated" people as Rembrandt, for example, they could suddenly draw with great facility. After they emerged from hypnosis, however, the subjects would remember nothing of their experience of "being" Rembrandt. Many would even scoff in disbelief upon being shown the artwork they had produced under hypnosis.[2]

Raikov demonstrated that the talents unleashed under hypnosis left significant residual effects after the sessions, even though the subjects no longer believed they were Rembrandt (or whoever). This residual effect made Raikov's method more than an experimental oddity—it was a practical tool for accelerative learning. Moreover, as we were to discover in Arlington, the Raikov Effect could be achieved without the aid of hypnosis.

Why Does It Work?

The Raikov Effect is little more than a modern-day resuscitation of an ancient and extremely powerful human practice.

Since prehistoric times, prophets, oracles, and tribal shamans have taken on the identity of gods, spirits, animals, and inanimate objects in order to gain knowledge. Twenty-thousand-year-old cave paintings of men wearing animal heads, found at places like Lascaux, almost certainly portray ecstatic rituals of the sort that anthropologists variously refer to as shape-shifting, spirit flight, or spirit possession.

Hunting cultures have traditionally organized themselves in clans that are identified with particular animal totems in the wilderness, for example, the eagle for the Eagle Clan or the bear for the Bear Clan. Members of the Bear Clan would hold rituals in which they imagined themselves becoming a bear, perhaps even wearing the hollowed-out head of a bear during the ritual and adopting a "bear" perspective in order to better understand the wilderness that was the source of their livelihood.

Even today, many believe that such experiences involve possession by real entities from another realm or dimension. In the modern world, such rituals can take the form of charismatic Christians speaking in tongues, Haitian dancers becoming voodoo gods, or New Age channelers speaking in the voices of extraterrestrials. Psychologists are more likely to ascribe such phenomena to a process of acute dissociation, in which one part of a person's mind takes on an independent personality of its own.

Whatever explanation you prefer, men and women have for thousands of years been sharing their bodies with other beings, real or imagined. There is overwhelming documentation that, while engaged in such trances, people exhibit skills, talents, knowledge, and even physical strength and dexterity unavailable to them in their normal lives.

The Merlin Method

One important use of the Raikov Effect is illustrated in a famous scene from the Broadway musical *Camelot* which portrays the British legend of King Arthur. In it, the wizard

Merlin transforms the young boy who will someday become King Arthur into various animals, in his imagination. While soaring aloft as a hawk, Arthur hears Merlin ask, "What does the hawk know that Arthur doesn't know?" Arthur looks down at the earth and realizes that, from the hawk's point of view, there are no borders in Britain. He resolves to forge a single nation from the patchwork of feuding tribes below.

Although fictional, this episode was inspired by a real tradition in Celtic folklore. The Druids of ancient Britain, of whom folklorists consider Merlin a late and only slightly Christianized example, were held to be accomplished shape-shifters, able to transform themselves and others into any form, animate or inanimate. The legendary Druid Mac Roth once flew into the heavens, wearing the headdress of a bird.[3] In an old Welsh epic, the bard Taliesin boasts, "I have been in many shapes . . . I have been a drop in the air; I have been a shining star . . . I have journeyed as an eagle . . . I have been a shield in fight; I have been the string of a harp . . . there is nothing in which I have not been."[4]

Force-Fitting?

It may be that such flights of imagination, if such they are, impart their wisdom through nothing more mysterious than DeBono's provocative operation or force-fitting effect, described in Chapter 7. In other words, they may inspire creative thinking simply by juxtaposing in the subject's mind a set of perceptions that do not ordinarily belong together. The subject's efforts to force a fit between these odd components yield a provocative new *gestalt*.

Such force-fitting played a major role in one productive brainstorming session held at Gillette Corporation in 1980. Executives were instructed to pretend that they were shafts of hair. While in their "hair" identities, they brainstormed what qualities would most please them in a shampoo. Some wanted a powerful cleanser that would root out dirt from the scalp. Others, fearing for their split ends, asked for a milder formula. In the end, the human "hair shafts" settled on a new shampoo that would automatically adjust to every hair need. Silkience, the product they invented, remains one of the leading shampoos on the market.[5]

MINDS WITHIN MINDS

Force-fitting is an important principle, but it hardly explains even half the phenomena associated with what I call Model Thinking. The human mind seems to have an almost infinite capacity for dissociation—splitting off discreet personalities within the same brain. For reasons that are still little understood, such dissociated personalities can exhibit powers and talents unknown to the host personality. It is almost as if the human brain were designed to house a multitude of different people.

Multiple Personalities

Sometimes people escape from childhood traumas by splitting into a whole spectrum of different, fully functioning identities, a condition known as Multiple Personality Disorder (MPD). So distinct are these personalities that MPD victims will have not only different handwriting styles, artistic talents, and knowledge of foreign languages, but even different allergies, illnesses, and reactions to drugs, depending upon which personality they are "using" at the moment. Scientists have confirmed that an MPD victim can exhibit different brain-wave patterns from one subpersonality to another—a feat almost as difficult as changing one's fingerprints.[6]

Talking Ghosts

One of the most striking dissociative phenomena I have run across came from an experiment conducted by Dr. Raymond A. Moody, Jr. Intrigued by worldwide folkloric traditions of mirrors and reflecting pools serving as gateways to the spirit world, Dr. Moody suggested to twenty-five experimental subjects that they could contact departed loved ones by gazing into a mirror. Twelve of the subjects reported "seeing" deceased persons in the mirror. A few even continued being haunted long after the experiment was over, when they were alone at home.

It will no doubt be debated in some quarters—and with perfect legitimacy—that these people saw real ghosts. Personally, I think it more likely that the apparitions represented dissociated elements of the subjects' own minds. This being the case, it is remarkable that several of the subjects reported engaging in lengthy and completely convincing conversations with the "ghosts" and that all alike reported the apparitions to be clear, solid, and completely lifelike—a striking observation coming from test subjects who were unhypnotized and who were all normal, sane people, none of whom held spiritualist or occult beliefs.[7]

At the very least, Dr. Moody's experiment has revealed an entirely new psychological phenomenon that may go far toward explaining the elves, fairies, gnomes, angels, goblins, and other strange but lifelike apparitions that have peopled folklore throughout history.

Lucid Dream Characters

Moody's "apparitions" may be linked to the phenomenon of lucid dreams, discussed in previous chapters. In the first historical mention of lucid dreams, by the Marquis Hervey de Saint-Denys in 1867, the Marquis claimed that he could, in his dreams, "call up the shades of the dead and also transform men and things according to my will."[8]

I suspect that the Marquis was not so much conjuring up departed spirits as creating dissociated dream characters so lifelike and independent in their actions that he believed them to be living beings. The peculiar powers of lucid dreaming have made possible a whole new field called intrapersonal psychotherapy, in which people work out their neuroses by resurrecting and confronting figures from their past, both living and dead.

Psychologist Paul Tholey of the University of Frankfurt, for example, was plagued by dreams in which his deceased father appeared in a menacing way. Tholey conjured up his father in a lucid dream and literally slugged it out with him, causing his father to devolve "into a more primitive creature, like an animal or a mummy."

"Whenever I won," Tholey reported, "I was overcome by a feeling of triumph."

Tholey also recommends a technique that strongly resembles Druidic shape-shifting—entering the body of another dream character. He points to the case of one teenage girl who, during a lucid dream, "entered the body" of a boy whose attention she had been soliciting in vain. Once inside her love object's "body," she began to see things from his point of view.

"I understood why he had been so reserved with me," she reported, "and I realized that he would never return my feelings." As a result, the girl was able to end a fruitless and disheartening infatuation.[9]

Borrowed Identities

Geniuses have long used the technique of symbolically borrowing others' identities as a tool for igniting creativity. Walt Disney, for example, drew inspiration from "becoming" the characters he created. "Mickey's voice was always done by Walt," remembers one Disney animator, "and he felt the lines and the situation so completely that he could not keep from acting out the gestures and even the body attitudes as he said the dialogue."[10]

George S. Patton thought himself reincarnated from great generals of the past. This odd belief may have catalyzed his eerie genius for applying the lessons of ancient battles to modern mechanized warfare.

Michelangelo imagined his statues as living beings, wholly formed inside the stone and awaiting only his hammer and chisel to free them. Of course, the unhewn statue lived only in Michelangelo's head, but envisioning it whole and alive somehow kindled Michelangelo's genius for "freeing" magnificent forms from the stone.

THE MASTER MIND PRINCIPLE

In his 1937 book *Think and Grow Rich,* the great motivational writer and educator Napoleon Hill pointed to teamwork between like minds as the most powerful engine of human achievement. Such "coordination of knowledge and effort . . . between two or more people" Hill called a Master Mind, a group mind that was greater than the sum of its parts.

Included in Hill's own Master Mind were a number of personages who existed only in his imagination. In *Think and Grow Rich,* Hill described a technique he used for resurrecting and conversing with great geniuses from history.

Each night before falling asleep, Hill would close his eyes and imagine himself to be in the company of nine "invisible counselors" modeled after his nine greatest heroes: Ralph Waldo Emerson, Thomas Paine, Thomas Edison, Charles Darwin, Abraham Lincoln, Luther Burbank, Napoleon Bonaparte, Henry Ford, and Andrew Carnegie. "My purpose," wrote Hill, "was to rebuild my own character so it would represent a composite of the characters of my imaginary counselors."

Hill would then address each member of his "cabinet" by name, asking them respectfully to bestow upon him some quality that he admired in that particular character. From Emerson, he requested "a marvelous understanding of nature"; from Napoleon, the "ability . . . to inspire men"; from Lincoln, a "keen sense of justice"; and so on.

To add substance to his characters, Hill studied their lives voraciously. After a few months of nightly "conversation" with his counselors, Hill was "astounded" to find them beginning to take on a life of their own. "Lincoln developed the habit of always being late," Hill recalls; ". . . Burbank and Paine often indulged in witty repartee which seemed, at times, to shock the other members of the cabinet."

The characters became so real that Hill actually suspended the meetings for several months, fearful that he was losing track of reality. Eventually, unable to resist the company of such august and stimulating figures, he returned to the practice. Hill maintained that he never regarded these "cabinet meetings" as anything other than "purely imaginary." Nonetheless, the wisdom he gained from them was utterly real.

"They have led me into glorious paths of adventure," he wrote, "rekindled an appreciation of true greatness, encouraged creative endeavor, and emboldened the expression of honest thought."[11]

THE SOCIAL BRAIN

It may be that dissociative phenomena have their root in the structure of the mind itself. In recent years, a growing number of researchers have embraced a concept that Dr. Michael Gazzaniga calls "the social brain." Gazzaniga believes that our minds consist of independent "modules," each one a "co-conscious" personality capable of independent thought, action, and even emotion.[12]

Perhaps the genius identities Dr. Raikov's subjects acquired during "artificial reincarnation" represented nothing more than artistic or intelligent modules of the social brain that for some reason had been previously repressed. Research on this question will no doubt continue for many years to come. In the meantime, we needn't wait for a theoretical explanation in order to avail ourselves of the Raikov Effect. Its astounding benefits are available now, through the simple methods outlined below.

THE MODEL THINKING APPROACH

Unfortunately, deep hypnosis such as that used by Dr. Vladimir Raikov is both difficult to attain and somewhat dangerous, especially in careless or ill-intentioned hands. For this reason, I was both astonished and overjoyed at our 1977 workshop to find that Mary, the precocious violinist discussed above, was able to achieve the Raikov Effect without the aid of a hypnotic trance.

Mary simply closed her eyes and imagined herself to be Jascha Heifetz, reveling in the sensory details of that experience. She was completely conscious the whole time and was never subjected to subconscious suggestion by any outside influence. Yet, as we have seen, Mary experienced a leap in accelerative learning as dramatic as any that Raikov had reported.

This laid the foundation for the technique I later developed called Model Thinking—a completely effective, nonhypnotic alternative to the Raikov method.

Putting on Heads

The essential technique of Model Thinking imitates prehistoric ritual. Stone Age shamans "became" a stag by putting on the hollowed-out head of a stag. We will do likewise, in our imaginations, "putting on the heads" of those characters whose consciousness we want to share.

Like Image Streaming itself, Putting on Heads improves with practice. Eventually, you will learn to draw insight from virtually any genius you select. You will gain more from this Borrowed Genius experience if you first spend some time reinforcing your Putting-on-Heads muscles by practicing the following exercise.

1. Think of a beautiful scene from your own experience. It can be a sunrise from a mountaintop, a sunset over reflecting waters, a wildflower-covered hillside in the sun and wind, the natural cathedral formed by tall trees in a deep woods—any experience that you can remember as having been especially beautiful.
2. Describe this scene as richly as possible, with your eyes closed, to a listener or to a tape recorder.
3. A lot of other scenes in your memory are superficially like this one, but this one is special. It is uniquely beautiful. In your description, focus on its uniquely beautiful qualities.
4. Imagine now that a companion is joining you in this experience. Your companion is extraordinarily perceptive—indeed, *the* most sensitive observer in the world. It might be someone you know, some historical figure, or even an entirely imaginary character. Let your unconscious decide who the character will be. Let your mind surprise you.
5. Solidify the character with rich, sensory description.
6. Imagine that the character is emanating a warm, welcoming presence. Revel in this aura of goodwill and describe it.
7. Now you are about to go "inside" this hypersensitive observer and find out what it's like to *be* him. To do so, imagine standing behind the observer at arm's length. Now let yourself waft inside the observer's body until your eyes align with the observer's own eyes, your ears with the observer's own ears.

   Another way to do this is to gently cup your hands over the observer's ears, gently lift the head off the shoulders, and fit it on over your own head like a helmet. Then pull on the rest of the body like a rubber suit.
8. Now you are seeing this same beautiful scene through the eyes of this keen observer. Experience all the observer's rich sensory impressions and sensitive awareness.

9. Certain things in the scene will seem different to you when you look through the observer's eyes. Focus on these differences and describe them in detail. Continue this for 3 to 5 minutes.
10. Take off the observer's head and place it back on his or her own shoulders.
11. Project a warm feeling of thanks to the observer. Let yourself be surprised, perhaps, at the warm feeling of thanks that comes back to you from the observer, thanking you for sharing such a remarkable experience.
12. Return to your immediate surroundings, with a full memory of all that has transpired and feeling remarkably refreshed.

The Borrowed Genius Procedure

The basic technique of Model Thinking is called Borrowed Genius. In this technique, you select a Model Genius and allow your consciousness to symbolically enter his or her body, thus gaining the benefit of that person's ingenious perspective. The procedure follows:

1. Select a particular skill that you are trying to acquire or a subject that you are studying. Choose a skill or subject that you can return to immediately after this exercise, in order to test the effectiveness of your Borrowed Genius experience through real-time practice.
2. Select a particular genius who exemplifies for you the highest achievement in the particular skill or area of study you have chosen. It can be a historical figure, someone you know, or even an imaginary character. This will be your Model Genius.
3. Close your eyes and imagine yourself in an exquisitely beautiful garden.
4. Describe the garden to your tape recorder in great sensory detail.

5. Make a slow, 360° turn, describing everything you see, in order to establish a strong sense of your position in space. Continue this for 4 to 6 minutes.
6. As always, let the Image Stream lead you. If the "garden" becomes something other than a garden, go with the flow. You can continue the Borrowed Genius exercise in whatever setting your unconscious mind chooses.
7. Imagine that your Model Genius has come to join you in the garden.
8. Begin describing the Model Genius in rich, sensory terms.
9. As you describe the Model Genius, imagine that your genius is projecting a warm, welcoming presence. Revel in his or her welcoming presence for 3 to 5 minutes.
10. Now that you have established strong neurological contact with the Model Genius, it's time to go for the *inside* view. You are about to find out what it's like to *be* a genius. Turn your genius around so you are facing his back. Stand your genius at arm's length from you. Now enter the Model Genius's body. There are two ways to do this. You can waft inside like a wraith, or you can gently grab the ears (as described above), lift off the head, and place it over your own head like a helmet and then pull the rest of the body around you like a rubber suit.
11. Align yourself with the genius's body. Put your eyes where his or her eyes are so you can see through the same eyes. Do likewise for the ears. Continue the procedure with all the other parts of the body.
12. Now look around the garden. You are seeing through the eyes of a genius. You will notice immediately that certain things look different. Describe those differences, from the genius's perspective. Continue for 4–6 minutes.
13. Now it's time to engage in that particular skill or activity at which your Model Genius is most expert. Suppose your genius is Rachmaninoff. Go to some part of the garden or other adjoining space where there is a piano.

Sit down at the piano in your guise as Rachmaninoff and start playing. Continue playing for 4 to 6 minutes.

14. While playing the piano (or performing the particular task, study, or skill you have chosen), describe in rich sensory detail everything you see, hear, taste, touch, and smell as Rachmaninoff. What are Rachmaninoff's characteristic gestures? What posture does he use? How does each separate part of his body feel as he plays at the height of his powers? What feelings and expressions does he have on his face? Focus intently on *body-related* feelings. Continue this for 3 to 6 minutes.
15. Go now to that moment in your Model Genius's life when he or she experienced his greatest illumination or epiphany, his peak "Aha!" moment, a time when everything came together and made perfect sense.
16. Describe as much as you can of that moment and of the perceptions and understandings that were part of it. Continue this step for 3 to 7 minutes.

Disengaging from Borrowed Genius

When you are ready to end your Borrowed Genius experience, simply do so this way:

1. Walk, in your mind, in front of a big, full-length mirror. See your Model Genius standing there facing you in the mirror. Now *abolish the mirror!* The mirror is gone, but the genius is still standing there facing you. You and the genius are no longer a single entity. You have returned to your own body.
2. Project a warm feeling of thanks to your Model Genius for allowing you the use of his body. Imagine your genius projecting thanks back to you for the privilege of sharing such a remarkable experience with you.
3. Now have your genius hand you a walkie-talkie. It is so small that you can easily slip it in your pocket. See that your genius is also pocketing a walkie-talkie. Let there

be a telepathic understanding between you that, by means of these walkie-talkies, you will always stay in touch, even after the Borrowed Genius experience has ended. Anywhere and anytime, you will be able to draw upon the talents and insights of your Model Genius.

4. Before you leave, your genius has something to say—he will impart to you some especially important insight about this experience. Listen closely. Report what your Model Genius says to your partner or tape recorder.

Debriefing from Borrowed Genius

As soon as possible after your Borrowed Genius experience, debrief yourself. With your eyes open but still speaking in the present tense, recount what happened during the experience. Describe it through a *different medium* than the one you used during the experience, either a tape recorder, another person, or even a notepad. Describe everything you experienced, especially the *differences* you noticed in the garden when you looked through your genius's eyes. Continue debriefing for 2 to 4 minutes, or longer if you are writing.

Borrowed Genius Follow-Up

As soon as possible after the Borrowed Genius experience, go and practice the skill you are working on. While practicing, try to develop the ability to slip easily back and forth between real-time practice and Borrowed Genius imagining.

For example, spend 10 to 30 minutes playing the piano and then 10 to 30 minutes *imagining* that you are playing the piano as Rachmaninoff. Try to alternate back and forth two to four times a day. As you steadily build confidence in your ability to slip back and forth between genius and nongenius states, at some point the two states will suddenly meld together in your mind. You will realize that you *are* that genius, playing the piano in real time. All the skill, talent, and perception of your imaginary Rachmaninoff will be yours.

When you use a Model Genius from history, it is, of course, beneficial to spend time at the library learning everything you can about the person, but it is not essential. Your unconscious database is so large and complete that it contains all the information you need in order to extract from any Model Genius the characteristics you desire.

YOU ARE THE GENIUS!

The geniuses you encounter in the above exercise are, of course, nothing more than dissociated elements of your own mind. Their mighty talents and subtle perceptions exist within you.

My experience has shown that Borrowed Genius is even more effective when the Model Genius you conjure up appears not as a different person but as an ingenious version of *yourself*. Unfortunately, most of us have already had our self-esteem so battered that we find it difficult to imagine ourselves as geniuses. For that reason, we suggest you practice, first, the "Borrowed Genius" procedure just given, to the point where you do find yourself clearly functioning at genius level in real time. Then, you might be ready for the more advanced form of the procedure, the Parallel World or Alternate Earth technique, described below.

PARALLEL WORLDS

Scientists tell us that our Milky Way galaxy contains billions of stars, many or even most of which likely have planets orbiting them. Despite this vast number, our galaxy is significantly smaller than our nearest major neighbor galaxy, Andromeda. Indeed, hundreds of millions of galaxies within range of our telescopes are thought to be far larger than our own.

In short, our universe is so large that it is statistically probable that somewhere out there are planets just like Earth, and that some of these planets may even have people on them who look just like you—indeed, who *are* you in every important respect.

Perhaps some of your cosmic counterparts had a kinder life on their parallel Earths. Maybe they never experienced the envy, the ridicule, and the poor teaching that held you back. Perhaps their parents were more understanding or their teachers more patient. Maybe they developed their talents to the fullest potential. On their own worlds, your counterparts may be geniuses of the highest order. Wouldn't you like to learn from them? Wouldn't you like to use their knowledge and experience to, in effect, go back for a second chance?

The Alternate Earth Procedure

1. Select some subject or skill you wish to develop.
2. Imagine that you are looking at the back of your own hand. In your mind's eye, picture the fingernails, the knuckles, the skin color, the texture, the fine hairs, the structure of tendons underlying the skin.
3. Now that you've strengthened neurological contact with your hand, imagine raising that hand and pushing an elevator button. Step back so you can see the whole elevator door. This is no ordinary elevator, actually, it is a Space/Time Transporter, capable of taking you to any planet in the universe. You are going to use it to visit an alternate Earth, where an alternate you has become a genius at the very skill you now wish to cultivate.
4. The elevator is coming from far away, so you have plenty of time to wait. While you wait, describe the elevator door in rich sensory detail (1–3 minutes).
5. The door opens. Step in.

6. Although the door was opaque on the outside, you see from the inside that it has a transparent panel you can see through. You will use this panel to navigate the elevator and stop it at the appropriate place.
7. To the side of the door, note the control panel, with pushbuttons set in it. There are a number of buttons, most of which you will not have to use in this exercise but are welcome to, on your own, on other occasions. One, for example, is labeled "Upwhen." That button will send you toward the future. The one below it is labeled "Downwhen." It will take you into the past. For this exercise, though, you will be using only the button labeled "Sidewhen." It will transport you to a parallel Earth, where one of your counterparts happens to be living. On the other side of this panel is a button labeled "Space," which could take you to that parallel Earth but would require you to cross a lot of space to get there. In one corner of the control panel is a little packet labeled "Disengage." Please disengage this packet from the corner of your elevator's control board and put it in your pocket. It is for emergencies (even though we've never had to use it). If you ever want to come back from one of these trips in a hurry, just slap that pocket and you'll be back in present time and space, on our own Earth Originale.
8. Now rest your finger lightly on the "Sidewhen" button, but *don't push it.* Program the elevator by *thinking* loudly but silently, with your finger still on the button, "Take me to a key point of experience on another, parallel Earth, where my counterpart, living there, happens to be that world's greatest genius in what I'm now seeking to learn."
9. Push the button. Allow yourself to feel the elevator moving. Perhaps you'll feel it moving up, or down, or front or back or to the side, or turning, or going in some direction you can't describe. Allow yourself to feel that movement strongly for a minute or so.
10. En route, you might catch glimpses of where you're headed, but the real action starts once you get there,

and the scene there may or may not correspond to what you glimpsed en route. Anticipate that the place where you will arrive is the one scene on that parallel Earth that is most relevent to the lesson you want to learn. Feel the elevator accelerating, picking up speed, going faster and faster.

11. Feel the elevator coming to a halt. The transparency in the door will flash a color; then the door will open. What color did the door flash? Remember that color so you can return to this site anytime.
12. Step out into the parallel world. Your counterpart self is there. Begin describing in rich sensory detail everything you see around you, including your counterpart self.
13. Now ease up behind your counterpart self and waft inside or put on your counterpart's head.
14. Align your eyes with your counterpart's eyes. Do the same for your ears. See and hear through the eyes and ears of your counterpart. Bring all of your mind and awareness into your counterpart's body and, for a while, *become* your counterpart.
15. Reach down through your counterpart's right arm with your own and pull on your counterpart's right hand like a glove. What does that look and feel like?
16. Do the same with your left arm and hand, pulling on your counterpart's left arm and hand.
17. Look around through your counterpart's eyes and notice what looks different from before. Describe the differences.
18. Begin performing the activity or skill at which your counterpart is a genius.
19. Note the body feelings that come with performing that activity—the characteristic posture, patterns of motion and gesture, patterns of body English, or muscle response—when that ingenious activity is going well.
20. Walk your counterpart over to a full-length mirror and look into it. Abolish the mirror. Now you are standing before your counterpart, face-to-face.

21. This is your chance to ask questions of your counterpart. Ask them silently but loudly in your mind. Listen closely for the answers.
22. Thank your counterpart warmly for the experience.
23. Return to the elevator. It deposits you back on our own Earth Originale feeling alert, refreshed, and charged with energy.

Learn the Steps Thoroughly

This Parallel World exercise is rather involved, so you must learn the steps well before beginning. Opening your eyes during the exercise to check instructions can break the spell. It might be helpful to think of the exercise in terms of a ten-step critical path:

1. Describe your *elevator door*.
2. Inside your elevator, rest your finger lightly on the *"Side-when"* button and tell your elevator where to take you.
3. Push the button and *feel* the motion.
4. You've arrived. What *color* do you see through the transparent panel?
5. Step forth into the *alternate Earth* and describe.
6. Find your *counterpart*. Describe.
7. Merge with your counterpart and describe the *changes* in your perception.
8. Perform the *target skill* in the guise of your counterpart.
9. Look into a full-length *mirror* and separate.
10. Engage in a *question-and-answer* session.

PUTTING ON HEADS FOR INSTANT ANSWERS

You can use Putting on Heads as a Thresholding technique while Image Streaming. If you have been Image Streaming for several minutes and no particular insight or "Aha!" experience has seized you, simply bring into your Image Stream a Model Genius whose insight you think will serve to elucidate the meaning.

Enter the Model Genius by the methods outlined above and spend some minutes describing the Image Stream through the eyes of that genius. Take special note of anything that seems *different* when viewed through the genius's eyes.

After you have withdrawn from the genius, either through the mirror technique or by simply removing the head and taking off the body suit, you can begin to put questions to your genius regarding the proper interpretation of your Image Stream.

AN ONGOING ADVENTURE

Putting on Heads is perhaps the most potent technique in this entire book. Once you master the basics, you will be astonished at the dramatic results you obtain.

Remember the baseball genius from the first chapter? He discovered his flyspeck principle by putting on the heads of various baseball heroes. Contained within you are extraordinary powers you've never even imagined, waiting to be given a face, a voice, and an identity so that they can communicate with you and impart their wisdom.

It is perhaps fitting that some of the most amazing results have come when this method was used for its oldest and most original purpose—taking on the personalities of animals in the wild. In one workshop, a participant "became" a bee buzzing by. A picture instantly flashed into his mind of the eave of a roof on the far side of a nearby building, with a hive tucked underneath. From the bee's point of view, this participant saw himself rapidly approaching the beehive from the air. You can imagine the curiosity of the other workshop participants. We all dropped what we were doing and trooped over to the location the man had described. Sure enough, we found a beehive exactly where he had said, swarming with bees. None of us had been aware of that beehive before, at least not consciously. Other workshop participants have had similarly remarkable experiences from putting on the heads of creatures such as dolphins and even shellfish.

More practical-minded readers may question the utility of such explorations. Yet they hint at an ever wider world of adventure that opens up when we learn—like the Druids of yore—to put on the heads of other beings. Modern science has scarcely begun to explore the mind's dissociative capacity. We can be certain only that this ancient faculty remains as potent for us as it was for our Stone Age ancestors and that its full potential has yet to be unleashed.

CHAPTER 9

TOTAL RECALL

One day, while reading Friedrich Nietzsche's *Thus Spake Zarathustra,* the great psychologist Carl Jung had a strange feeling that he had read a certain passage before. In fact, it appeared to be an almost verbatim copy of a lengthy episode Jung had read years before in a high seas adventure story published in 1835—half a century before Nietzsche wrote *Thus Spake Zarathustra*.

Jung wrote to Nietzsche's sister and learned from her that Nietzsche had indeed read that very same adventure tale when he was eleven years old. "I think, from the context," Jung later wrote, "it is inconceivable that Nietzsche had any idea that he was plagiarizing this story. I believe that fifty years later it had unexpectedly slipped into focus in his conscious mind."[1]

UNLIMITED MEMORY

Nietzsche had apparently experienced *cryptomnesia,* or concealed recollection. This occurs when forgotten memories seep into a person's consciousness without that person being aware of it. Cryptomnesia was likely responsible for putting ex-Beatle George Harrison in the embarrassing position of having to explain why his 1970 hit song "My Sweet Lord" bore such a strong resemblance to the 1963 Chiffons' tune "He's So Fine."

Although sometimes embarrassing, spectacular cryptomnesic feats such as Nietzsche's raise provocative questions about the power of human memory. If an entire passage of a book can be absorbed at age eleven and then be recited intact decades later, what else might be stored in our brains? And how can we retrieve it?

THE MEMORY BARRIER

"Education," Albert Einstein once remarked, "is that which remains, after one has forgotten everything he learned in school."[2]

Who among us has not, at one time or another, looked back ruefully at the years we spent cramming for tests in such mysterious subjects as physics, chemistry, trigonometry, history, French, and Spanish? Who has not marveled at the thought that once upon a time we actually knew how to read a periodic table, work with quadratic equations, calculate the trajectories of falling bodies, name the key battles of the War of 1812, and conjugate verbs in French—and then promptly forgot these skills?

By the time we enter the job market, most of us are lucky if we remember our multiplication tables. Any American who can stumble through a conversation in a foreign language is regarded with awe. As our desks at work pile high with unread trade journals, computer manuals, reports, magazines, and books of all kinds, few of us can escape a sinking feeling of *déjà vu*. In the workplace, as in school, we feel ourselves swamped with a mass of data far too large for our memories to hold.

METASKILLS

On an occasion when Einstein was asked the speed of sound, he replied, "I don't know. I don't crowd my memory with facts that I can easily find in an encyclopedia."[3] Contrary to popular myth, great geniuses are not necessarily great mnemonists or memorizers. They do, however, know how to find out what they want, when they want it. I call this a *metaskill*—a fundamental skill from which other skills derive.

Strong evidence suggests that all of us possess photographic memory to some degree. Only a few people, however, seem gifted with the ability to retrieve and examine their mental "photographs" at will. In this chapter, you will learn about PhotoReading, a metaskill that will help you retrieve far more of your hidden memories than you have hitherto thought possible.

The Retraining Frenzy

All adult readers of this book are facing a possible crisis in their careers. According to economist Paul Zane Pilzer, author of *Unlimited Wealth* and *God Wants You to Be Rich,* "The task that we perform today will be technologically obsolete tomorrow."[4]

In 1958, a single silicon chip could hold only ten electronic components. By 1972, that number had grown to a thousand. Today's state-of-the-art Pentium chip contains 3 million transistors crammed into a 1-inch square.[5] The rate of technological change continues to accelerate. Whole industries are being wiped out in the blink of an eye. Compact disks obliterated the multibillion-dollar vinyl record industry in less than ten years. Fuel injection and other digital systems made conventional car repair obsolete. The same story is being repeated in every industry. Success in today's turbulent new economy is won not by those who have learned a particular skill but by those who can learn new skills quickly.

"We used to have a society where we would educate you and then you'd do the same thing for the rest of your life," says Pilzer. "That doesn't work anymore. The median career length is now less than six years. Half the population changes *careers*—not just jobs—in six years or less."[6]

Learn New Tasks Faster

As never before, adults today must think of themselves as perpetual students, ready at all times to take on additional course work. The most competitive workers will be those who have mastered the metaskills—what Pilzer calls "basic skills"—enabling them to learn new tasks faster.

As an office manager, for example, you might find that a new computer system doubles your ability to promulgate, organize, and disseminate documents. "If your basic skills are highest in your department," writes Pilzer in *Unlimited Wealth*, "you will probably be the first one to learn the new system. . . . By the time your coworkers master the new technology, there will likely be an even better system available, starting the whole process all over again."[7]

ACCELERATED LEARNING—YOUR COMPETITIVE EDGE

For many, a society based on constant education is a frightful prospect. How will we find time for all that retraining while supporting ourselves and our families? Accelerated learning is one answer. Only by sharply improving our metaskills can we reduce the time demands of retraining.

Project Renaissance is currently engaged in a study attempting to compress two years or more of college into an 8-week course of summer school. We intend to enroll 200 students in the 1996 pilot program, called the Project for Accelerated Academic Learning (PAAL). After completing

PAAL, students will go on to one of five participating universities where they will be tested for possible admission at the junior-year level.

The methods of PAAL can be easily adapted to compressing virtually any on-the-job training into a fraction of the time. They include all the Project Renaissance techniques described in this book, such as the various applications of Image Streaming and Model Thinking, as well as others we discuss later, such as Freenoting and Buzzgrouping. They also include PhotoReading.

MEMORY—THE ULTIMATE METASKILL

The roster of traditional metaskills includes such abilities as reading, writing, speaking, reasoning, and calculating. Your ability to learn more specialized skills—such as using a computer—depends in large measure on how well you have grasped these fundamentals.

Yet even these metaskills grow from a deeper and more powerful antecedent—your ability to retain and retrieve data from your memory. Enhancing this ultimate metaskill is therefore a major goal of PAAL and a backbone of most other accelerative learning programs.

PHOTOREADING

Among the most potent of our PAAL methods is a technique called PhotoReading. PhotoReaders not only vastly increase the speed and quantity of their reading, but they also retain and retrieve useful data at many times their ordinary capacity.

Like so many geniuses, Paul Scheele, the inventor of PhotoReading, was a poor student in school. Reading proved to be an especially slow and cumbersome task for him. Speed-reading courses helped Scheele raise his speed from 170 to 5,000 words per minute at 70-percent comprehension. Yet Scheele found the process of zooming through page after page of type to be exhausting and enervating.

Through his work as an accelerative learning consultant, Scheele heard about a speed-reading instructor in Phoenix, Arizona, who had made a remarkable discovery. In order to learn "eye-fixation patterns," students would practice scanning pages of a book at high speed, upside down. On a hunch, the instructor administered a comprehension test following one such exercise. To his amazement, the students attained record scores. They had somehow managed to absorb the material subliminally.

Subliminal Learning

The peculiar power of subliminal learning has long been recognized. As far back as 1916, researcher L. L. Thurstone discovered that sailors who were taught Morse code in their sleep learned it 3 weeks faster than those who were taught normally.[8]

In 1954, a movie theater in Fort Lee, New Jersey, experimented for 6 weeks with high-speed advertisements flashed on the screen between movie frames. Messages such as "Hungry? Eat popcorn" and "Drink Coca-Cola" would blink onscreen for 40 milliseconds at a time—far too quickly to be noticed by the audience. Yet, during the 6 weeks of the test, popcorn sales rose 57.7 percent and Coca-Cola sales 18.1 percent.[9] Public disclosure of the experiment prompted widespread demands to outlaw subliminal advertising. However, the outcry soon died down and the practice was never banned. Subliminal advertising remains to this day a multibillion-dollar industry.[10]

Obviously, a simple message saying "Hungry? Eat Popcorn" would hardly have affected the audience had it been flashed on the screen for a normal 3 seconds or more. People would have read it, yawned, and ignored it. Only when it bypassed their conscious minds did the message become compelling. Information perceived subconsciously impacts the brain far more powerfully than conscious information.

The IDS Breakthrough

In 1985, Scheele was hired to develop a speed-reading application for IDS/American Express. Inspired by the results from Tucson, Scheele reasoned that everyone might have a latent eidetic, or photographic memory, that could somehow be tapped. He developed a method for this purpose that he called PhotoReading.

The underlying principle was the same as that used in the Fort Lee movie theater in 1954. Scheele's plan was to expose readers to information far too quickly for their conscious minds to process it. PhotoReaders would glance at each page for only a second or two. After finishing a book at this speed, they would have no more conscious awareness of what they had read than had the patrons in the Fort Lee theater. Yet Scheele believed that the "mental photographs" of the book could later be activated and used.

He was proved right. IDS employees who learned Scheele's method reported astonishing improvements in the quantity of material they were able to consume and remember. A standard speed-reading test showed Scheele that he could Photoread a book at 68,000 words per minute yet display 74 percent comprehension.[11]

A Demonstration

In May 1986, Scheele's company, Learning Strategies Corporation, was licensed by the Minnesota Department of Education to teach the new method as a vocational curriculum. Some skeptical professors at one Minnesota college tried to prevent Scheele from teaching his PhotoReading course, arguing that it couldn't possibly work. In response, Scheele arranged a demonstration. In the presence of the professors, one of Scheele's colleagues PhotoRead a volume of U.S. patent law.

"Afterwards," writes Scheele, "he scored 75 percent comprehension. In addition, he drew approximations of six patent illustrations and correctly identified their numeric sequence."

An Evolutionary Leap

PhotoReading appears to be a natural step forward in the evolution of human reading skills. In ancient and medieval times, people commonly read out loud. Most found it difficult to read without at least moving their lips. Today, almost every literate person can read silently. Yet, we still tend to subvocalize while we read. Our brains and nervous systems unconsciously go through the motions of speaking. This habit slows us down tremendously, since it reduces our reading speed to the speed of our tongues.

PhotoReading frees us from the need to subvocalize, allowing us to read at brain speed rather than tongue speed. When we were just learning our ABC's, we had to sound out each letter before recognizing a word. Later, we learned to recognize whole words at a glance, or even whole sentences and paragraphs, if we took a conventional speed-reading course.

PhotoReading takes this process one step further, allowing us to absorb two whole pages at a time. At first, it seems impossible, because you are reading far too quickly to recognize and vocalize each word in turn. Once you shed the habit of subvocalization, however, you will soon find that it is far more comfortable to let your brain process a text at its natural speed—very, very, fast!

THE PHOTOREADING PROCEDURE

A full explication of Scheele's technique can be found in his book, *The PhotoReading Whole Mind System,* which can be ordered from Learning Strategies Corporation in Wayzata, Minnesota (call 612-475-2250 or 800-735-8273). The following pages present a basic outline of the procedure.

Step 1: Enter the PhotoReading State

Place the book or other reading material before you. Close your eyes and become aware of every part of your body. Sit up straight. Put both feet on the floor. Breathe deeply and evenly. Relax.

Now imagine that you are hovering just above and behind yourself. Open your eyes and look down at the pages, imagining that you are reading over the top of your own head. When you read, you will notice your two hands holding the book on either side. The technique of peering over your head has actually widened your visual perspective, allowing you to take in both facing pages at once.

At this point, your brain waves should have slowed to a leisurely 8 to 12 cycles per second—the so-called Alpha State (your normal waking Beta State is 12 to 23 cycles per second). Psychologists have found that this relaxed but alert state of mind is ideal for learning.

Step 2: State Your Purpose for Reading

Before you start reading, silently state your purpose for reading this particular material. Francis Bacon once said, "Some books are to be tasted, others to be swallowed, and some few to be chewed and digested." Paul Scheele put it more bluntly: "Some things are worth reading in great detail. Others are not worth reading at all."

Most of us never plan our reading. We pick up books and magazines at random and simply start reading. This behavior is terribly inefficient. Reading a book can take days, a long article more than an hour. Such time-consuming tasks should be managed as carefully as any other important job you do.

If you want to savor Shakespeare's sonnets, you should probably read them the old-fashioned way, "chewing" and "digesting" every word. But if you subscribe to an online service and have avoided reading the 250-page user's manual for six months, you may prefer to "swallow" it whole through PhotoReading. Be realistic. Ask yourself before starting:

- *What do I expect to gain from reading this material?* Will the book teach you to use your new computer, explore a favorite hobby in greater depth, speak a foreign language, or gain insight into a current social issue? Determine in advance exactly how you expect to be made more capable or better informed by reading the material.
- *What level of detail do I want?* Do you need to learn and remember every fact in the book, the major points of each chapter, or just one or two key facts or techniques? Decide in advance.
- *How much time am I willing to commit right now to achieving my purpose?* After answering the first two questions, you may decide that the book is not worth your time at all. Perhaps it's worth only 5 minutes. You be the judge.

Step 3: Preview

The greater feel you develop beforehand for the book or article as a whole, the easier you will find it to PhotoRead. First, *survey the book's overall structure.* Read the Table of Contents; the titles, subtitles, and subheadings of all the chapters; the index; the text on the front and back cover; and any boldfaced or italicized print. Also read any boxes, diagrams, or charts. This should give you an excellent feel for the book's structure and content.

Next *identify the key trigger words,* that is, words that jump out at you as you scan the book. The author uses them again and again because they represent key facets of the book's theme. In *The Einstein Factor,* such terms as

"Image Streaming," "Squelcher," and "Feedback Loop" will have jumped out at you. Find out precisely what the trigger words mean, and you will understand the book.

Finally, *do a mini-inventory.* Have your priorities changed during the previewing process? In many cases, you will find that after completing the preview you have obtained everything you need from the book.

Previewing Is Optional Readers may find that they can PhotoRead quite well without previewing. Indeed, there may even be some advantage in doing so. Priming the brain beforehand with too much conscious information about the book can sometimes block more subtle insights from the subconscious. For this reason, I have been working with Paul Scheele to devise less intrusive methods of previewing. In the spirit of a true scientist, Paul has been remarkably receptive to my suggestions and eager to test them experimentally. Early test results have been encouraging, and we will likely use a revised previewing method in the PAAL program (in which Scheele and I are collaborating). Until we iron out all the kinks in the new method, however, readers should choose (with probably comparable success rates) between the original method herein described and no preview at all.

Step 4: Approach the Theta State

Before PhotoReading, you want to achieve profound relaxation that opens up your right brain and gives you access to unconscious memory. That means descending into a deep Alpha State very close to the Theta State, which lies between 4 and 8 cycles per second. The Theta State is ideal for working with mental imagery and other right-brain activities.

Close your eyes and breathe deeply, letting anxiety flow out with each breath and tranquillity flow in. Each time you exhale, think of the word *relax*. The first few times you PhotoRead, you might want to lie down first and relax each major muscle group of your body, one at a time, while repeating the word *relax* each time you exhale. Be careful not to fall asleep. Our body tends to link the Theta State with bedtime. Eventually, you will gain the knack of entering this state without such elaborate preparations. With practice, a couple of deep, slow breaths will serve as an adequate trigger.

An alternative is to engage in a 5- to 10-minute session of Image Streaming before you PhotoRead. It will achieve the same effect.

Step 5: Attain PhotoFocus

When you focus precisely on single words or sentences, your left brain switches on and your right brain switches off. So much for your Theta State! When you are PhotoReading, *never* focus on specific words. Instead, concentrate your attention on the white space *around* the letters. Imagine yourself looking over the top of your head as you gaze at the entire page at once, focusing only on the white space. You will see both pages at once, as well as your two hands holding the book on either side. The page will take on an almost three-dimensional appearance. When you gain that sense of depth, you have attained PhotoFocus. This way of seeing is quite similar to the semifocused vision you use when viewing the new stereoptic Magic Eye pictures.

Step 6: PhotoRead

Now start turning the pages, allowing 1 or 2 seconds per page. Remain relaxed at all times. Let your mind stay as blank as possible. Avoid negative thoughts or gently bounce away from them if they occur. Especially avoid generating such self-fulfilling prophecies as: "This isn't going to work." One way to keep your mind relaxed and positive is to chant to the rhythm of the page turning. You might chant "Relax . . . Re-lax . . ." or "Keep the state . . . See the page . . ." Do not try to read the pages at all, but be sure to look at each pair of pages for a second or two.

Continue PhotoReading to the end of the book. When you are finished, don't test yourself by trying to recall what you have just PhotoRead. You will only get discouraged. You need to follow the activation procedures described below in order to retrieve any of the information.

Step 7: Sleep on It

Give the information time to incubate in your mind. If you're in a hurry, you can begin activation after about 20 minutes. However, it is best to wait a full 24 hours. Researchers at the Weizmann Institute in Rehovot, Israel, have discovered that learned skills consolidate in the brain during REM sleep. Performance of certain tasks actually improves the day *after* the new skill is learned, even when there has been no practice in between.[12] A similar effect seems to occur with PhotoReading.

Step 8: Activate

Sometimes information that you have PhotoRead will be activated spontaneously, without your conscious awareness. A professor at a state university in Minnesota once PhotoRead two books in preparation for a speech he was giving. Then he slept on it. During his sleep, the professor dreamt about giving his speech. He suddenly woke up and scribbled down everything he could remember from his dream. In the morning, when he looked at his notes, this professor saw that his speech was 90 percent complete.

Another PhotoReader prepared for a trip to Sweden by PhotoReading a Swedish dictionary several times. At a restaurant, he suddenly found himself ordering a meal in Swedish.

Such stories are fun but rare. Most of the time, you need to activate your information "manually."

Activation Step 1: Probe Your Mind Start by *asking yourself questions.* "What is important to me in that book I just read?" you might ask. "What do I need to know to perform well on the next test, to write my report, or to contribute in the next meeting?" Don't expect instant answers. This sort of mindprobing is meant only to get your unconscious thinking about the best ways to resurrect the stored information. The important things are to stay relaxed, remain confident, and stay curious.

Activation Step 2: Scan and Dip Your mind probe should have made you curious about certain sections or passages in the book. Turn to these sections. Now scan rapidly down the center of each page. If any word or phrase catches your eye, dip into the page and read a line or two until you are satisfied. Then resume your scan pattern. Ration each dip to a paragraph or two of any article, or a page or two of any book. Follow your hunches, not your logic. Allow your subconscious to direct where you will dip into the text, even if the passage seems irrelevant to your purpose. If you mistrust your initial hunches or want greater thoroughness, scan and dip through the entire book, from the first to the last page.

Reading experts estimate that 90 percent of any text is filler. Scanning and dipping will lead you directly to the substantive 10 percent.

CULTIVATE THE PHOTO-FEEL

The early effects of PhotoReading will not be obvious to most people. The effect comes first in the form of vague feelings and hunches that you might easily ignore if you are not looking for them. Becoming an expert PhotoReader involves developing a feel for or alertness to these subtle inner signals. One of the best ways to develop this heightened awareness is through Image Streaming. Your daily 10-minute pratice of Image Streaming trains you to notice and react to precisely the sort of inner signals that comprise the Photo-Feel, whether they surface as images, words, or subtle urges. (Paul Scheele now asks all his trainers to Image Stream as preparation for teaching their PhotoReading classes.)

After a frustrating meeting spent trying to understand his attorneys' legalese, one businessman went to a bookstore and PhotoRead every book he could find on the subject for which he was seeking advice. He was about to leave the store when he felt a strong urge to go back and scan the shelves. The man walked right over to a particular book and opened it to the exact page he needed.

An engineer at a power plant was amazed to find himself at a meeting discoursing on a subject in which he had little expertise. Only later did he remember having PhotoRead a large stack of trade journals in his office. A quick scan and dip through the stack revealed an article he had completely forgotten, but which had provided the background for his stellar performance at the meeting.

THE WHOLE-MIND EXPERIENCE

The metaskill of PhotoReading has applications far beyond digesting books and journals. Practiced PhotoReaders have used the technique to enter a Whole-Mind state, giving them far easier access to right-brain processing in their daily activities.

One jeweler dreaded an upcoming trade show. Each year, he would march laboriously up and down the aisles, searching every booth for the stones he wanted to buy. The ordeal usually lasted about five days.

This time, however, he tried PhotoReading the exhibits. First, he stood back to get a wide-angle view of the entire auditorium. Then he walked briskly down the aisles, allowing his eyes to scan the booths in PhotoFocus. After probing his mind about what sort of stones he felt he needed, the dealer walked once more down the aisles, scanning and dipping as he went and stopping only at those booths that "called out" to him. The jeweler completed his mission in a record 2 hours.

USE THE SKILL UNTIL IT FEELS REAL

Your biggest obstacle to success in PhotoReading will be the temptation to give up because you don't believe it's working. Like every other skill you have learned in this book, PhotoReading improves with practice. As you get better at PhotoReading, you will gain a deeper sense of how it works and how to trigger its most potent effects.

The best test of PhotoReading's effectiveness is to start using it in as many different situations as possible. It won't be long before you compile enough of your own success stories to become more than convinced.

THE MEMORY EDGE

The clearest mark of genius, in the popular mind, is the mastery of left-brain skills such as advanced mathematics, fluency in foreign languages, and the wielding of an impressive vocabulary. Yet these achievements are based on little more than rote memorization. PhotoReading can accelerate your progress in every one of these areas.

Try PhotoReading a dictionary several times, whether in English or some foreign tongue. This exercise will greatly enhance your vocabulary. In tandem with formal study, PhotoReading will accelerate your progress in almost any area many times over. It is a metaskill with few equals.

CHAPTER 10

THE SOCRATIC EFFECT

During the Korean War, Americans were horrified to see U.S. prisoners of war reading anti-American statements before the camera and professing belief in Communism. How had their Chinese captors persuaded them so easily? It turned out that the Chinese had used *hsi nao*—an expression that literally means "wash brain." American journalist Edward Hunter translated it "brainwashing," and the name has stuck to this day.[1]

What is truly remarkable about Chinese brainwashing is that it requires no torture or physical coercion. Instead, captives are persuaded by small degrees to express alien opinions of their own free will. They may first be asked to do nothing more than write a list of everything that might be good about Communism. Later on, they may be asked—under the guise of fairmindedness—to write a list of everything wrong with the American position. The process proceeds in relentless, minute increments until at last a captive is converted.

SELF-EXPRESSION

The power of brainwashing lies in the fact that the subject is not expressing views that someone else has put in his mouth. On the contrary, he knows full well that he has generated these views himself. Ultimately, the growing gap between his self-image as a loyal American and the fact of his increasingly long list of disloyal confessions causes what psychologist Leon Festinger called a cognitive dissonance—a jarring inconsistency in a person's thoughts, feelings, and behavior.[2]

According to Festinger, people have a strong tendency to eliminate cognitive dissonance. If you force yourself to smile and act cheerful when you are sad, for example, you will eventually react to the cognitive dissonance by either giving up the pretense of happiness or actually becoming happy. You cannot maintain the two conflicting states for very long—nor could those hapless American servicemen in Chinese prison camps.

SOCRATIC METHOD

Nowadays, we think of education as a process of cramming information into a student's head. But the Latin word *educare* means literally "to draw out." In ancient times, it was the educator's job to draw out the students' own subtle perceptions and insights (see Figure 10.1).

This education followed Socratic Method. Although Socrates didn't invent it, he greatly popularized the technique. In Socratic Method, the teacher asks a series of acute questions, forcing students to examine, defend, and describe their perceptions and ideas.

Among the many benefits of Socratic Method is that it causes students to reach their own insights and express them in their own words. As the Chinese brainwashers

Fig 10.1 The word *educate* comes from Latin *educare,* "to draw out." The ancients believed that wisdom came from within. In Classical Athens, great teachers like Socrates drew out their students' subtle perceptions through Socratic questioning.

would no doubt agree, there is no surer way to make a lasting impact.

Socratic Method benefits the teacher as much as the student. Indeed, the ancient Greeks formed schools as much for the benefit of the learned teachers as for their students. Through teaching, the leading thinkers of Classical Greece ensured that they would have audiences before which to air their ideas and perceptions. These teachers, called Sophists, would then return the favor by drawing out the perceptions of their listeners or students, through Socratic questioning. Both sides availed themselves of a powerful feedback loop that spurred their intellect and widened their perceptions. A remnant of this system survives in the popular aphorism "If you wish to learn a subject, teach it."

THE DEMISE OF EDUCATION

Socratic Method allowed Classical Athens—a city-state of fewer than 100,000 people—to achieve feats of art and learning that still inspire awe 2,400 years later. It remained the basis of Western pedagogy until 150 years ago. Then the huge numbers of pupils flooding into American public schools led teachers to abandon Socratic Method. Faced with classrooms of forty to sixty restless children, teachers no longer had the leisure to question one or two pupils at a time in the drawn-out Socratic style.

At that point, we stopped educating and started teaching. Today's didactic teaching methods presume that each student is a *tabula rasa*—a blank slate—that the teacher must fill through lecturing, as if pouring water into an empty glass (see Figure 10.2). The profound wisdom naturally inherent in each student is no longer drawn forth, as it

Fig 10.2 Modern educators see the student as a blank slate, or "tabula rasa," waiting to be filled with facts from without.

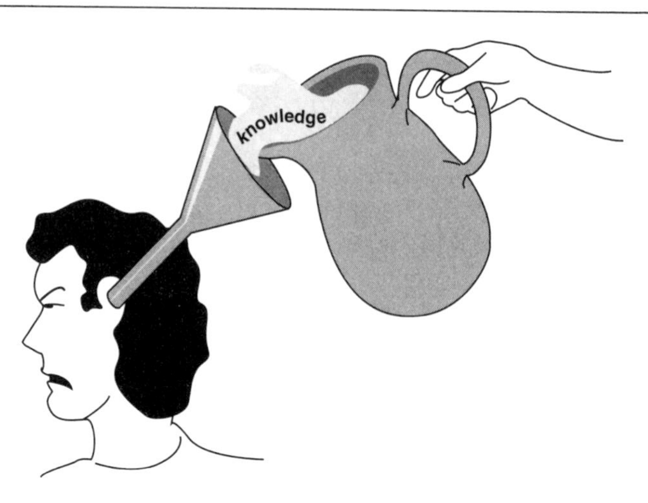

was in Periclean Athens. Socratic Method survives today only in the close one-on-one relationships that develop over many years between professors and graduate students. Law schools claim to use Socratic Method, but they use it in a severely distorted fashion in which the threat of immediate dismissal for unsatisfactory answers terrorizes students into paying rigid attention. It is a method more inquisitorial than Socratic.

Only now are we beginning to realize the heavy price we paid when we abandoned education in favor of teaching. Einstein's previously quoted remark has taken on a bitter new meaning for modern educators: "Education is that which remains after one has forgotten everything he learned in school."[3]

THE SELF-EXPRESSIVE INSTINCT

Who has not sat through a meeting, champing at the bit to voice some opinion? No matter how eloquent or persuasive the other speakers, we hardly hear them, so absorbed are we in rehearsing over and over our own planned speech. Even after we speak, we continue to play and replay our own words in our mind, relishing every nuance and syllable while the ongoing meeting proceeds around us in a formless buzz. When the meeting is over, guess whose words we remember best?

What is perhaps most frustrating is that the more interesting the meeting is, the more we are aroused to speak and the less we hear what is going on!

This instinctive practice is looked upon as discourteous and self-centered, as indeed are most instinctive habits when they are left unchecked. How many well-meaning self-improvement books have adjured their readers to "Shut your mouth and *listen* to people!" Yet the advice falls on deaf ears. The need to express ourselves at all costs is hard-wired into our brains as deeply as our drive for food or sex.

The Traffic Jam in Your Articulariae

What we think of as human consciousness is largely seated in the *articulariae*, those brain structures that govern self-expression. Among them are Broca's Speech Area, which enables us to talk, and Wernicke's Area, which helps us read and comprehend others' speech. More physical modes of expression, such as writing, sketching, painting, and dancing, also make extensive use of the Supplemental Motor Area.

The articulariae pluck particular perceptions from the boiling sea of impressions in our mind and give them concrete form so that we and others can make sense of them. For millions of years, the natural direction of flow through the articulariae has been relentlessly *outward.*

Unfortunately, 99 percent of modern schooling is an attempt to jam information *into* the brain through those very same word-processing structures—a situation exacerbated by the ever-diminishing funding in our schools for nonverbal instruction such as art and music. Over time, the resultant traffic jam in our articulariae blunts and diminishes our ability to perceive and learn.

Mental Judo

Practitioners of the martial art of judo, instead of blocking or countering their opponents' attacks, divert the force of their opponents' own moves against them. We can use this highly efficient principle to break the traffic jam in our articulariae.

It will always be an uphill—and mostly fruitless—battle to try to bludgeon and shame people into "shutting up and listening." Even if we succeed, how much a person will retain of such passively imbibed information is questionable. Imagine how far the Chinese brainwashers would have gotten had they sat all the prisoners down in a classroom, taped their mouths shut, and harangued them for hours with Marxist doctrine!

Project Renaissance has developed a highly effective technique of mental judo that gives full vent to self-expression during any meeting, lecture, or classroom situation, while at the same time allowing instruction to enter the brain subliminally. This technique, which has become a cornerstone of our PAAL curriculum, is called Freenoting.

FREENOTING

Most book lovers abhor the idea of writing in any book. Yet, paradoxically, the most valuable antique tomes are those bearing the marginal scribblings of some noted genius. Great minds have a peculiar compulsion to comment on the spot whenever some passage they are reading provokes their imagination. Such marginal notes offer historians a fascinating glimpse into the thought processes of some of history's most profound thinkers.

Freenoting is an extension of this time-honored compulsion. It simply means writing down, on the spot, whatever thought pops into your head as you listen to a class, meeting, or lecture or even as you read a book. The difference between Freenoting and ordinary note taking is that you make no attempt whatsoever to capture the facts that the lecturer is imparting. On the contrary, you allow your scribbles to range as far and wide from the speaker's topic as your mind carries you, writing constantly and rapidly so that it is virtually impossible for you to follow the lecture. This method ensures that the lecture material will enter your mind subliminally, rather than consciously.

Like PhotoReaders, Freenoters have found that they retain far more of their Freenoted lectures than they do of lectures to which they tried their best to pay attention. Moreover, their scribblings are found to contain brilliant insights that integrate the subject material far more intimately and practically with their own private interests and situations than would normally be the case.

Seeing Is Believing

As with so many accelerative learning techniques, in which we are asked to absorb information in radically different ways, Freenoting arouses skepticism in many beginners. The idea of reading or sitting through a lecture or meeting and *not paying attention* seems a formula for futility, if not disaster. For that reason, I was very pleased recently to receive the following letter from Tony Brigman of Grand Prairie, Texas:

> I have to admit that I was skeptical about your claims that freenoting ideas into a cassette recorder as they come into your mind while reading written material can actually help you remember with greater comprehension what you have been reading. So, this morning, I decided to put it to the test.
>
> With my trusty tape recorder in hand, I picked up a chapter to read from the scriptures. (How about that for the ultimate comprehension test!) As I began to read, fresh thoughts, associations, ideas, arguments, etc. would come to mind about what I was reading, so I would simply record them on tape and then proceed with my reading. I noticed that the more I commented, *the more rapidly additional insights and thoughts and associations began to come into my mind.*
>
> When I finished reading the chapter, I closed the book and then made an attempt to recall the content of what I had been reading. I was ASTONISHED. I remembered more than I can ever recall remembering from reading scriptures. As you know, some of the scriptures are not exactly easy reading. . . .
>
> How is it possible to have GREATER RECALL of specific material when you are simply recording thoughts, ideas and arguments that come to mind *about* the material, and NOT concentrating on MEMORIZING the material? All I know is it happened to me this morning. There is merit to what you're saying, Win. I had to test it for myself. I look forward to more pleasant discoveries.

As Mr. Brigman learned, seeing is believing. The best argument I can make for the efficacy of Freenoting is to try it yourself.

The Freenoting Procedure

1. Bring a regular notepad and tape recorder to the lecture or meeting you are Freenoting.
2. Tape the lecture as you Freenote. You can use this tape later for a scan and dip review.
3. Write quickly and continuously, allowing your mind to roam free, though staying more or less on the topic of the lecture. You can also Freenote directly onto tape, but this may annoy your neighbors in a lecture or classroom setting.
4. Forget self-consciousness. No one's looking over your shoulder. Don't censor. Write down bad ideas as well as good ones, and don't cross out anything. In most cases, you won't be able to recognize your best ideas as such until *after* you've written them down or recorded them on tape.
5. Don't stop or hesitate. Your goal is to force a flow of data outward through your articulariae. The flow must be so rapid that most of the lecture, to enter your mind, has to get through subliminally.
6. Use doodles, cartoons, and diagrams. Any involvement of right-brain functions will increase your neurological contact with the subject matter and enhance your learning. Aim to stay more or less within the context of what you assume to be the topic (I say "assume" because you may not be entirely sure what the topic is, especially if you're Freenoting rapidly and effectively).
7. Seek deeper patterns. Probe for the underlying connections between the subject matter of the lecture and human knowledge in general. Don't be afraid to invent far-reaching metaphysical systems, on the spot, that purport to tie together all events in the universe. This indicates

that your unconscious is struggling to make the subject matter relevant to your worldview—a key step in learning any subject profoundly.

The More Bizarre, the Better

Throughout history, great minds have used the power of association to jog their memories. This means linking the information you want to remember with some imagined object, color, or sing-song rhyme. Freenoting will automatically increase your memory of the lecture by associating key points with your own expressive thoughts. The crazier, funnier, and more bizarre those thoughts are, the more memorable they will be.

In a chemistry lecture, you might imagine those atoms that combine easily with hydrogen to be covered with fur in which the little hydrogen atoms tend to get caught. Picture the strange, varied coats of fur that would distinguish each type of hydrogen-trapping atom. Oxygen, for example, might have very wet looking fur.

In general, think of the lecturer as playing straight man to your comedian. Poke fun at the lecture. Think of the subject matter as being straitlaced and conservative, and strive to spice it up in your Freenoting.

The Freenoting Gestalt

In Chapter 5, we discussed how a jumble of seemingly random thoughts congeals in a gestalt far greater than the sum of its parts. Freenoting works in a similar manner. When the lecture is complete, you will find that your notes contain more and better information than was imparted in the entire lecture. Your greater consciousness has gone beyond the surface facts and grasped the underlying pattern behind the speaker's words. From that framework, you have unconsciously built original structures of thought and fresh observations.

So profoundly does your unconscious fathom the deep structure of the lecture that Freenoters are often startled to hear the lecturer saying things they had written just moments before!

Freenote This Book

You can test the effectiveness of Freenoting right now, using this book as your first "lecturer." When you are Freenoting a book, simply stop at regular intervals and rapidly scribble your thoughts (or speak them into a tape recorder) for 5 to 15 minutes without pause.

The best time to start Freenoting is when a thought or inspiration pops into your head. However, you can also stop yourself at random, at the end of chapters, or after certain regular time intervals. No matter when you stop, some interesting associations will always lurk in your head, diverging creatively from the straight subject matter of the book. In your Freenoting, take those stray thoughts as far as they will carry you and then return to your reading.

I predict that if you use Freenoting as you read the remaining five chapters of this book you will be amazed at how many more fresh, original insights you take away from those five chapters than you did from the first ten. No doubt you had just as many insights while reading Chapters 1 through 10, but some of them were lost because you failed to note them down.

In Freenoting *The Einstein Factor,* you might well come up with original techniques of your own for enhancing learning and intelligence. Your methods may bear little resemblance to mine but work just as well for you. Even if your Freenotes read like rough copies of my thoughts, you can trust that your greater consciousness has imbued them with a perspective that is uniquely your own and that tailors the information to your special needs.

Project Renaissance would be most interested in hearing of any Freenoting insights you receive while reading this book. Your ideas could very well help us develop new theories and methods for accelerative learning. Please write to us in care of the publisher.

LEARNING IS A CREATIVE ACT

When you eat a carrot or a slab of steak, you cannot expect it to replenish your body until after your digestive tract has broken it down into a different form. Likewise, information that enters your brain must be transformed before your mind can use it. There is no "magic pill" of knowledge that you can swallow whole. The act of learning is an act of creation on the part of the learner. The data from a book or lecture comprises only the raw material. You have not truly learned it until you have turned it into something new.

THE PRINCIPLE OF ARTICULATION

In Chapter 3, we discussed the Principle of Description, an idea that actually arose from an earlier metaprinciple I call the Principle of Articulation. This principle holds that *the more you express or articulate a given perception, the more you will perceive and understand of that and related perceptions.* The Principle of Articulation provides the underlying rationale of both Socratic Method and its corollary, Freenoting. The sheer act of persistently expressing our thoughts on some subject causes us to learn more about that subject, even when no new information has been provided from without.

What enhances our knowledge, in such cases, is not the addition of outside facts, but rather our own perceptions *about* our perceptions, feeding back into our minds in an ever-growing snowball effect. I cannot emphasize too greatly the awesome—and all-too-often neglected—power of the

Principle of Articulation to enhance performance in virtually every field of human endeavor.

Musical Feedback

Suppose we applied this snowball effect to areas outside book learning. Quite by accident, I discovered one way to do this almost thirty years ago, long before I had formulated my theory of the Einstein Factor.

Between 1965 and 1969, I picked up the habit of tinkering on the piano with a tape recorder running. I had no musical training, and probably even less natural talent. Yet I found that if I tinkered without inhibition or self-consciousness, simply pretending or acting as if I were playing a real piece, some portions of the ensuing cacophony would actually resemble music. As I listened to these tapes, I found that I tended to screen out the discordant parts while the pleasing parts stuck in my head. This feedback loop reinforced in me the behavior of producing pleasing sounds. Within a short time, the tendency to emit discordant noise had entirely dropped out of my consciousness, and I found myself playing lengthy, highly musical pieces quite spontaneously. All of these pieces had a clear beginning, middle, and end. In this manner, I soon "composed" dozens of highly listenable pieces.

The Composer Within

As we discussed in the section about Putting on Heads in Chapter 8, it is a firm dictum of accelerative learning that all possible talents and behaviors are contained, to some degree, within each of us and can be brought out through selective reinforcement of those qualities.

Each of us possesses a powerful instinct for musical composition. The music-like themes that emerged from my spontaneous keyboard banging sprang from deep wells of inspiration in my subconscious mind. Even after the pressures of work, children, dogs, and other domestic mayhem forced me to suspend my composing for nearly thirteen years, I found that when I finally resumed composing, my unconscious picked up right where it had left off, extrapolating from the same familiar, but evolving, patterns and themes that apparently comprise my unique musical "language."

IMPROVITAPING

The beauty of the Improvitaping Technique, as I subsequently named my improvisational recording, is that it serves the beginner and the professional musician equally well. In the beginner, it builds confidence and creates an instinctive feel for musical performance and composition. In the professional, it unlocks creative inhibitions that may have been stymied by excessively rigid training or by a debilitating awe of the composing process.

Like Image Streaming, the Improvitaping Technique harnesses the power of chaotic forces. From the turbulent interference between your spontaneous musical expression and your own attentive feedback, "standing waves" suddenly arise in elegant new forms.

The Improvitaping Procedure

1. Select an instrument with which you would like to work. It doesn't matter whether you have been trained to play it or not.
2. Play your chosen instrument for 30 minutes per day while recording on blank tape.
3. Avoid trying to play familiar themes written by others, but do your best to simulate the feel of original music, no matter how awful it sounds.

4. In some ways, this step is the hardest part. For 60 minutes per day, play back the tapes you made. Just listening to the sounds you made can be an excruciating experience at first. Listen attentively for 30 minutes, then play the tape in the background while you pursue other activities for the next 30 minutes.
5. Pay special attention to the parts of your tape that seem most pleasing. Through the First Law of Behavior, those themes that work for you are reinforced and will recur and evolve in your playing. Those parts that bore you or offend the ear will gradually drop out.
6. Continue your daily practice for at least 2 weeks. The Wince-and-Groan phase won't last more than 3 to 5 days. During this time, your persistence, faith, and good humor will be tested to their limits as you recoil from the sound of your own taped improvisations.
7. Toward the end of your first week, good thematic material will begin showing up here and there. You can mine this ore from the tapes and, by more conventional methods, compose them into pieces. I call this phase the Melody Mine. During this time, the more discordant elements of your playing will drop out through the snowball effect of your constant feedback and reinforcement. In the early stages, try to avoid compulsive repetition of one or two simple melodies or routines. These could easily become a crutch that blocks out deeper and more complex expression.
8. After the first week, you should enter the Whole-Piece phase. The music flows, developing its own intrinsic logic and form. Your Improvitapes emerge as complete, original compositions filled with rich surprises. You will be truly amazed at how melodious and pleasing your playing has become.

We have found that variations on the Improvitaping Technique work just as well in other artistic media, such as painting or sketching. A good source of techniques for applying the Socratic Effect to drawing and other arts is *Drawing on the Inventive Mind by* Jon Pearson (self published, 1992, PO Box 25367, Los Angeles, CA 90025, Tel: (310) 312-9022). Pearson's book is geared for children and is particularly suitable for use in schools.

THE POWER OF SELF-EXPRESSION

In general, what is expressed *by* the learner is a hundred times more productive learning than what is expressed *to* the learner—a statement to which Socrates himself would have heartily assented.

You learned in Chapter 3 to open the pathways to your Image Stream by increasing your neurological contact with the imagery. Likewise, all our experience on the frontiers of accelerated learning suggests that the best way to learn is to increase neurological contact with the target subject. In pursuing this goal, you will find few methods to be more potent than the ever-growing snowball effect of Socratic self-expression.

CHAPTER 11

THE OXYGEN FACTOR

Dr. Yoshiro NakaMats is perhaps the world's foremost living inventor. Among his 2,356 patented creations are such pillars of modern technology as the floppy disk, the hard disk, and the digital watch face.

NakaMats' thinking methods are as original as his inventions. According to *Success* magazine, when NakaMats wishes to brainstorm, he plunges into the swimming pool and swims underwater as long as he can. While swimming underwater, he scribbles ideas on a special Plexiglas slate he invented for the purpose. Only when he can't hold his breath another second does NakaMats finally resurface. He claims to get his best ideas through this method, which he calls "swim till almost die."[1,2]

Wherever NakaMats goes, journalists flock to "the Thomas Edison of Japan," having learned that they can rely on him for outrageous quotes and zany behavior. Yet there is more to NakaMats' underwater swimming technique than mere eccentricity. It has a firm basis in the physiology of the brain.

THE POWER OF PNEUMA

Nowadays, we take for granted that air is composed of free-floating atoms in the form of gases. But the ancient Greeks could only guess at the nature of that mysterious substance that sighed invisibly through the trees and filled their lungs with breath. They called it *pneuma*—spirit.[3] To the Greeks, the lung, or *pneumon,* was the organ of their bodies that drew in spirit from the surrounding air. The Romans likewise referred to breath as *spiritus.* To this day, we say that we *expire* when we give up our last breath, while *inspiration* means literally the drawing of air into our lungs.

THE ROLE OF OXYGEN IN BRAIN FUNCTION

Like so many ancient beliefs, those concerning *pneuma* and *spiritus* contained more than a grain of truth. Our brains, in which reside all that we think of as the human spirit, are totally dependent upon oxygen. Fully one-third of all the oxygen used in our bodies goes directly to the brain. Evidence suggests that the more oxygen we receive, the better our brains function. Einstein's brain, for example, possibly received more oxygen (and other blood-borne nutrients) than most. Marian Diamond found that rats raised in a highly stimulating environment had enlarged capillaries and a higher density of glial cells, which are believed to act as mediators between the neurons and blood vessels of the brain.[4] As noted in Chapter 1, Diamond found a similarly high density of glial cells in Einstein's brain.

Dean Falk, an anthropologist at the State University of New York (SUNY) at Albany, has even proposed that increased blood flow through the brain may have caused our apelike ancestors to evolve human intelligence. She points out that early hominids living in the African grasslands 2 to 3 million years ago evolved a "radiator" system of dense bunches of veins to cool their crania in the hot African sun. Falk believes that this intricate blood network made it possible for the hominids' brains to grow larger.[5]

Underwater Swimming

In the summer of 1959, I attended summer school in an attempt to make up for some serious deficits in my grades. Because my afternoons were free and the school had a pool, I spent several hours each afternoon swimming. I found that, for some reason, swimming underwater felt better to me than normal swimming. An unusually large amount of my time was therefore spent each day holding my breath for extended periods, as much as 4½ minutes at a time.

A remarkable thing occurred. Despite my lackluster academic record, my poor study habits, and the demanding course load I had taken on, my grades suddenly shot through the roof. On test after test, I found myself scoring 100. By the end of the summer, I had literally gone from the very bottom of my class to the top.

Only many years later did I gain an insight into how this might have happened. I was attending a lecture by the late Dr. Robert Doman, medical director of Philadelphia's Institutes for the Achievement of Human Potential (IAHP), founded by his brother Dr. Glenn Doman. During the lecture, Dr. Doman explained that whenever the carbon dioxide (CO_2) content of the blood increases, our bodies interpret it to mean that our oxygen supply is being cut off. In response, the carotid arteries that carry blood to our heads open wide and allow more blood to flow through them, drenching the brain in an unusually rich flow of oxygenated blood. In more rigorous times, those of our ancestors who weren't equipped with this safety device didn't live long enough to *become* our ancestors. Today we can harness this primitive reflex to push ourselves forward on the evolutionary path to greater intelligence.

OXYGENATING YOUR BRAIN

Increasing the flow of oxygen to your brain will accomplish two things. First, it will activate areas of your brain that are usually idle from lack of blood. Second, it will slow down the constant die-off of brain cells.

Inside the skull, your carotid arteries branch into smaller and more numerous arteries, fanning out in a fantastically intricate network of lacy capillaries. This dense network is designed to reach into every crease and corner of your brain in order to feed as many neurons as possible. Yet, inevitably, some cells will be less well supplied than others. These tend to be the cells you use the least and are also the first to die off.

After the age of thirty, the brain's circulatory system becomes less and less efficient. At least 35,000 brain cells will die every day—200 in the time it took to read this far in the chapter. Over the next week, almost a million more will likely die. Since humans have at least 100 billion brain cells, this rate of loss is hardly noticeable. Yet, it adds up over the years. Moreover, as the brain's circulatory system continues to deteriorate, the neurons that die will tend more and more to be active, useful cells rather than idle ones.

You can arrest and even reverse this process by increasing your cerebral blood flow. As more blood flows into the brain, an equal amount flows out through the veins. This increased drainage has the added benefit of washing away toxins and wastes that interfere with brain function.

Masking

In general, the carotid arteries tend to overreact. They admit far more extra blood than is needed to compensate for tiny increases in CO_2. For this reason, Dr. Doman suggested, an effective method for oxygenating the brain may be to induce small increases in the CO_2 content of our blood. One technique Dr. Doman recommended for this purpose is called masking.

Masking simply means breathing into a constricted space for a few minutes (the IAHP uses a special mask for the purpose). Each exhalation contains a little less oxygen and a little more CO_2. Masking for a minute or so will cut your oxygen intake only a tiny bit, but it will cause the carotid valves to open so wide that they are virtually flooding the brain with oxygen and nutrients.

After the first time you mask, the blood flow will soon return to normal. However, if you mask regularly for 30 second intervals every 30 minutes of your waking day and maintain that regimen for 2 to 3 weeks, you will train your carotid arteries to admit more blood on a continuous basis. Dr. Glenn Doman and his colleagues at the IAHP have come to regard masking as a potent method for improving brain function. Millions of their patients have used it safely over the years with powerful effect.

However, as the IAHP warns, masking may be hazardous. You should never mask without consulting a doctor first. The specialists at the IAHP in Philadelphia, for example, never advise masking until after they have assembled a detailed medical history of the patient.

The Diving Response

Any kind of vigorous aerobic exercise, such as jogging or stair-stepping, will increase your CO_2 level and improve circulation to your brain. Underwater swimming, however, is far more effective, in my opinion, than either aerobic exercise or masking.

Underwater swimming stimulates what marine biologists call the diving response. When we dive, the body increases blood flow not only to the brain but to every other major organ as well. This response is common to all mammals and may partially explain why whales and dolphins—perhaps the champion breath holders of all time—have evolved brains as complex and powerful as our own.

In the 1930s, British marine biologist Alister Hardy proposed that our ancestors may have lived mainly in the water. The aquatic ape theory (AAT) best explains why we lost our fur coat; why we have a layer of fat beneath our skin like whales, dolphins, seals, and hippopotami; why we have conscious control of our breathing (other land mammals don't); why we stand upright (to keep our heads above water in shallow marshes); and why we have sebaceous glands, which secrete waterproof oils on our skin. If our ancestors really were aquatic apes, their deep-diving habits may largely explain why they developed big brains. We can emulate that same evolutionary path through underwater swimming.[6]

Perhaps just as important, underwater swimming is fun, making it more likely that you will persist long enough to attain long-term effects. Readers who have access to a swimming pool should strive to spend as much time as possible swimming underwater. Don't push yourself. Build up your endurance gradually. **As with all other techniques in this chapter, you should consult a physician before starting.**

The Gravity Position

Gravity is nearly as effective as CO_2 in increasing blood flow. I recommend that every reader try the Gravity Position while practicing your 10- to 15-minute daily quota of Image Streaming.

Lie flat on your back on the floor, with no pillow. Rest your feet and lower legs over the seat of a chair or sofa. Make sure your legs are supported up to the knee so you don't stress them, but not so high on the knee as to impede circulation (see Figure 11.1). Loosen any tight clothes. Once you've settled in and are comfortable, luxuriate in several deep sighs. Finally, close your eyes and begin Image Streaming with a partner or tape recorder.

We have found that the Gravity Position lends a unique intensity to the Image Stream and that a brain enriched with extra blood flow is quicker to grasp the "Aha!" insights.

When you rise from the Gravity Position, especially the first time, you should get up slowly to give your circulatory

Fig. 11.1 Image Streaming is more effective when the brain is enriched with blood. This can be achieved by the Gravity Position.

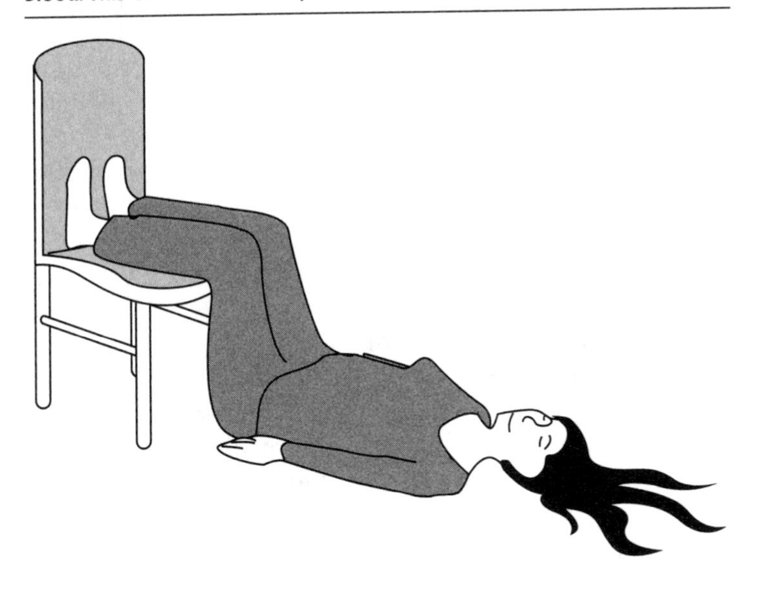

system a chance to adjust. The initial logginess you feel will turn quickly to freshness and clarity. Because the body's natural biological clock tends to slow us down in the afternoon, many cultures take a siesta during that time. However, 10 to 15 minutes of Image Streaming in the Gravity Position will add remarkable zest to your afternoons.

BREATHING AND AWARENESS

At this moment, as you start to read these words, *you are holding your breath.* Gotcha!

Whenever you turn your attention to some new stimulus—say, a new sentence—you hold your breath. You can't help it. Eventually, you let yourself breathe again. But the moment you resume breathing, a mental chain reaction automatically kicks in whereby you release your attention from the previous stimulus and turn it toward something new. The moment your attention locks onto something else, your breathing freezes once more. Note that I said the moment your *attention* locks onto something, not the moment your eyes fix upon it.

We also hold our breath at the start of a sustained physical effort, such as lifting a heavy load. Some people, while fixed intently on some task that demands attention (such as fixing a car engine), will hold their breath so long they actually become dizzy.

Now that I've told you this, you will probably deliberately try to read the next few sentences without holding your breath. It isn't hard to do. You *can* override this instinctive behavior if you try, but only for a while. Soon you will forget and return to your normal routine.

The Attention Pacemaker

In the 1980s, former Secretary of State George Schultz proved surprisingly ineffective during his first years under President Ronald Reagan. Despite his high intelligence and sterling track record, he seemed unable to formulate a coherent foreign policy and floundered desperately in cabinet meetings, unable to effectively defend his positions. It's not hard to see why. Videotapes of his TV interviews during those years show that he was always very short of breath and often had to pant before he could finish a sentence. The resultant damage to his attention span seriously impaired his ability to perform.

Try it yourself. After returning out of breath from your next jog, try focusing on some intellectual activity, such as reading a book. You won't be able to concentrate until after your breathing has settled down. Your breath is, in effect, a pacemaker for your attention. If you take short breaths, you will tend to have short bursts of attention and to speak in short sentences. Deep, full breaths will enable you to speak in longer, more complex sentences and to form deeper thoughts.

Underwater swimming is the best remedy for over-short breath. The longer you practice it, the longer you will be able to sustain a single breath—and a single thought. Who knows . . . Perhaps the Iran-Contra scandal would not have gotten so out of hand had George Schultz spent an hour a day in the swimming pool!

BREATH, THE RHYTHM OF LIFE

Few major bodily functions fail to respond sharply to changes in the depth or speed of our breathing. The ebb and flow of oxygen through our lungs truly defines the rhythm of our being. With over 40 percent of the body's energy being burned in the brain, that organ is affected more than any other by the flow of oxygen—the principle fuel for the body's metabolism.

Expanding or impeding the blood flow to your brain gives you incredible leverage over brain function, either for better or for worse. I urge all readers to drink deeply of the *pneuma* around you. You don't have to swim, as Dr. NakaMats does, until you "almost die," but I do suggest you start visiting a conveniently located swimming pool.

CHAPTER 12

WORKING IN GROUPS

When he was nine years old, an Oglala Sioux boy became delirious with an illness. He dreamt that four herds of mighty horses came galloping in from the four corners of the earth and that the Six Grandfathers of the World—the guardian spirits of his tribe—appeared in a cloud and gave him important messages for his people.

The boy recovered from his illness but kept the dream to himself. When he was sixteen, the boy was smitten by an intense fear of thunder, whose clamor reminded him of the thundering hooves in his dream. A medicine man warned the boy that he should tell his vision to the tribe. Only when the message was delivered would his fear go away.

The old man's advice proved correct. When the boy unburdened himself of the vision, his fear vanished. Moreover, the tribe decided to dramatize his vision in a new ceremony in which horses were made to converge from four different directions. Afterwards, the entire tribe felt stronger and better, and many sicknesses were healed. The boy, whose name was Black Elk, went on to become a powerful medicine man in his own right. He recorded all these events in the book *Black Elk Speaks.*[1]

DISTANT THUNDER

Many people today are like that fearful young boy, trembling at the sound of distant thunder. Fear and pessimism seem to dominate the public mind. Opinion polls show that most Americans expect their personal freedom to diminish in the years ahead. We imagine that our children will grow up less free, less healthy, less safe, and less prosperous, inhabiting a starving, overcrowded world strangling in toxic fumes and poisoned waters. No wonder psychologists report an unprecedented rise in clinical depression among all age groups—a phenomenon Marilyn Ferguson dubbed "The Great Depression" in her book *The Aquarian Conspiracy.*

Yet, as Ferguson points out, the great psychiatrist Karl Menninger believed that mental disturbances often herald dramatic breakthroughs to higher levels of thought. Can it be that our current malaise is but a precursor to positive change? Is the turmoil in our hearts simply a healthy discontent with an old way of life whose time is fast drawing to an end? I believe that it is.

SHARE THE VISION

As in the case of Black Elk, I believe that a big part of the cure for our public depression is to emerge from our isolation, to go into the community, and to share our private visions with the "tribe."

Today there are more doctors, scientists, Ph.D.s, college graduates, and literate people walking around than at any time in human history. By one estimate, 90 percent of all the scientists who ever lived are alive today. No previous generation had access to such massive brainpower.

We need to start networking—to go out and begin talking to one another and to share our views, our opinions, and our dreams. Only in this way can we pool the massive intellectual resources now at our disposal. I believe that the collective genius of mankind is crying out today through dreams, intuitions, and Image Streams, just as the Six Grandfathers cried out to Black Elk in his vision. Our subconscious minds already know how to solve every problem that weighs on our hearts. Like the Oglala Sioux, we need to create group rituals that allow the unconscious to speak.

Group Image Streaming

After fifteen years of conducting Image Streaming seminars, I have become convinced that group sessions are, by far, the most potent way to consult our inner genius. We humans are, at root, social creatures. Our brains seem to function best when they are working with other brains toward some common purpose. Image Streaming is rewarding in solo, but its full potential comes out only in groups. Many times I and my seminar participants have been delighted at the joy, mirth, and penetrating insight that come bubbling forth from a well-attuned group of Image Streamers.

Readers who are serious about following the program set forth in *The Einstein Factor* should seek like-minded people in their communities. Keep in mind that, in the age of the Internet, your community can include people anywhere in the world. An enthusiastic support network will intensify your feedback loop and motivate you to stay with the program. Weekly or monthly Image Streaming parties (think-tanks) at your house will forge wonderful bonds between friends and infuse your life with adventure.

THE GROUP METHOD

There are many ways to practice Group Image Streaming, but all require a leader or moderator who keeps everyone focused, instructs newcomers, keeps track of the time, and talks the group through each procedure. If the group continues to meet through the years, the leadership should rotate from time to time so that everyone has a chance to experience the added insights that result from leadership responsibility.

During timed procedures, I generally strike some metal object on a glass partially filled with water to signal when participants should begin or end a particular activity. A Chinese gong or virtually any other pleasant but penetrating sound will serve just as well. This is called the "chime."

Below I offer a time-tested script for leading an Image Streaming group. You can either use this format live or pre-read it onto an audiotape for later use in solo or group sessions. In such cases, the audiotape takes on the role of leader, so no one is left out of the action.

Prep the Group

1. Make sure the room or space you are using will be undisturbed for at least 30 minutes. Disconnect all phones within hearing distance.
2. Ask each participant to choose a partner, so everyone can work in pairs. If there is an odd number of participants, one group can have three people.
3. Ask the partners to sit very close together so they can hear each other easily, even during times when everyone in the room will be talking at once.
4. Explain the "chime" system to the participants. A single chime indicates that the particular exercise in progress is coming to a close in 30 seconds. When participants hear it, they should continue what they are doing but be ready in 30 seconds to stop talking and listen to the next instruction. Three or more chimes mark the end of whatever step of that exercise is in progress. Upon hearing them, participants should stop talking immediately, even in midsentence or midword, and listen for the next instruction. Imagers should keep their eyes closed, however, and continue imaging while they listen, to keep the experience itself alive and moving forward.

Group Script for Image Streaming

1. "Please pair off as partners and decide between yourselves which one is to be the Spotter and which one the Imager. I'll tell you in a minute what Spotters and Imagers do. Before I do that, please decide between you and your partner now who will be a Spotter and who will be an Imager."

In a group of three, two should act as Spotters. Allow 2 minutes for people to pair up and choose roles.

2. "Thank you. Does everyone have a partner? Does everyone know whether they are Spotters or Imagers? Good. Now, the Imager is the one who will be experiencing a spontaneous Image Stream in just a few moments, when we get under way. The Spotter will help the Imager tune into the imagery. I'll explain in a minute how we do that. Everyone will get a chance to be both Spotter and Imager. After we do this process for a few minutes, I'm going to ask everyone to reverse roles. Okay? Now let's get started.

3. "According to these instructions I'm reading from [you can use a copy of *The Einstein Factor* as a prop], every one of us has a constant stream of images going through our heads, which we usually don't even notice. This is called the Image Stream. These images contain a lot of unconscious wisdom, in symbolic form, much like our dreams. But, with practice, we can learn to access the Image Stream when we are wide awake and learn to get very vivid images.

   "Many people have been conditioned to believe that they can't see mental images. In fact, everyone has a constant flow of imagery. We just have to learn how to notice it. If I ask you to think of the Taj Mahal and start describing it, you will certainly see an image of the Taj Mahal in your mind's eye. Even without such prompts, simply closing your eyes and waiting will result in some of your stronger images jumping out and seizing your attention. That's where your Spotter comes in. When a stronger-than-usual mental picture enters your mind's eye, you respond automatically by holding your breath or allowing your eyeballs to move beneath your closed lids, tracking whatever object you're seeing. When the Spotter sees these attention cues, he or she will ask you, 'What are you seeing right now?' This will help you notice the images in your mind at that moment, if you need some help in becoming aware of them. Now let's get ready to Image Stream."

4. "Imagers, please close your eyes now and keep them closed until I later ask you to change roles with your partner. Have all Imagers closed their eyes? Good. Now, Spotters, look directly into your partners' faces and watch attentively. When you see your partners' eyes move beneath the eyelids, that usually means they are tracking some image in their Image Stream. The moment you see that, ask your partner, 'What are you seeing right now?' Don't say it yet. First, I want to do a practice run, so you Spotters will know what to look for. Imagers, please move your closed eyes under the lids back and forth one or two times. Do it now. Spotters, please watch while the Imagers move their eyes. There's a difference between mere eyelid movements and the actual eyeball moving beneath the lids. Eyelid movements are of no use to us. We want you to notice only when the eyeball itself moves. Some people's eyelids are so thick that it's hard to see the eyeball movement. In those cases, just notice the slight sideways movement of the eyelashes. It's very distinctive. Let's try it once more to make sure all the Spotters know what to look for. Imagers, move your eyeballs now. Okay? Is everyone clear? Great. Let's start Image Streaming."

5. "Imagers, for those of you who are beginners, don't try to look deliberately for images. That deliberate effort tends to get in the way of the spontaneous responses we want. The most important thing is to just relax. We're going to do a technique called Velvety-Smooth Breathing that will help you relax. Imagine you are feeling a long, pleasant, very smooth strip of velvet. Starting right now, try to make your breathing feel as good as possible. Breathe slowly, deeply, and velvety-smoothly. Stroke your breath through your body like you'd stroke a piece of velvet and make that feel as smooth as possible. There should be no pauses between your in-breath and your out-breath. When you finish breathing in, you should go smoothly right into your next out breath. Just make it all one smooth, continuous b-r-e-a-t-h-e and keep it going. Work at making each in-breath of that one smooth, continuous b-r-e-a-t-h-e feel as good as possible. With each out-breath, see how

many of your tensions and anxieties you can simply let go. Let them flow right out with that breath. And keep on making that in-breath feel as good as possible. It's all part of one deep, rich, smooth, continuous b-r-e-a-t-h-e. Now if anyone does start to see an image, start describing the dickens out of that image to your partner. Don't wait for your partner to ask you. But if you don't notice an image, just keep on breathing and relaxing. And keep on velvety-breathing and relaxing while I instruct your partner."

6. "Now, Spotters, any time your partner pauses in his or her breathing, that's an attention-cue. It means that your partner has just noticed something happening in the Image Stream. Whenever you see that, or see your partner's eyeballs moving, gently ask your partner, 'What was in your awareness just then?' or 'What are you seeing now?' The first time you ask this, your partner may not be able to tell you—maybe even the tenth time. But sooner or later, your partner will discover that, yes, maybe he did see two orange dots off on the right or, yes, maybe she had for a moment been looking at a tree limb, or a daisy moving in the breeze, or some other sort of impression. Spotters, when in doubt as to whether or not you've actually seen a cue, go ahead and ask anyway. Chances are pretty good you've caught a response."

7. "Imagers, when any image comes, you should describe the dickens out of it to your partner, no matter how slight or ephemeral or even silly that image might seem to you. And Spotters, keep encouraging your partner to describe any image. Your coaching will help. Even if it's just a flash memory, Imagers, and you saw it only for a second and you have to describe it from memory, go ahead and describe it, using the present tense as if you are still seeing it in the here and now. The very fact of your describing it will cause the image to take form once again in your imagination. Describe it as vividly as you can, using all five senses. Describe it in such rich detail, with so much color and power, that your Spotter won't be able to help seeing what you're seeing. The more you describe, the more images will come up for you to describe. Just let the images flow freely, from one image to another."

8. "Imagers, are you still breathing? In continuous, smooth breaths, without any pauses? Good. Spotters, are you looking for attention-cues

and asking questions? Good. Imagers, keep breathing smoothly, pleasantly, and continuously, and whenever anything at all comes up in your Image Stream, without waiting to be asked, describe the dickens out of it. Start now!"

Allow 3 to 5 minutes, depending on how the exercise is going. Check around to make sure that each pair is working effectively. It's better to undermanage than to overmanage. Don't intrude on anyone's process. A good way for you to spend this time is to "look in" silently on your own Image Stream for a few minutes while the group is working. Thirty seconds before ending this step, sound your chime once and make the following announcement.

9. "That's the half-minute warning. Continue as you are, but be ready 30 seconds from now for further instructions."

Thirty seconds later, end this phase of the exercise by sounding the chime three or more times.

10. "Good. Thank you! Now, please, all Spotters become Imagers and all Imagers become Spotters. Has everyone switched roles? Proceed as before, in your new roles. Begin now!"

After getting the process under way again, time it as before, giving the 30-second one-chime warning. Finally, end this step, as you ended the last, by sounding the chime three or more times.

Co-Tripping

If all pairs seem to be receiving spontaneous images fairly easily and switching roles without problems, you might want to move on to the co-tripping option you first learned in Chapter 3. If not everyone wants to co-trip or you sense restlessness, skip down instead to the close-out.

"Since everyone is doing so well, let's try a new exercise called co-tripping. Imagers, please continue being Imagers. All Spotters, however, should now become Imagers, too, and begin Image Streaming. Each of you tell your ongoing experience to the other, a few sentences at a time, informally, on your own, rather than awaiting formal turns. When one of you pauses for breath, the other should resume describing his or her own Image Stream. Then, as that one pauses for breath, the other should rush in, so that no air time is wasted between you. Both of you should go full-tilt, describing your own images while hearing the other's. Start now!"

Give the Imagers 3 to 5 minutes of co-tripping, wrapping up with the usual 30-second warning chime and the final three or more chimes to signal the end of the exercise.

The Close-Out

Whether or not you pursue the co-tripping option, the close-out is the same. When you are ready to end the Image Streaming, say the following:

"Very good. Excellent, everyone. Now, as easily as that, and with full recall of everything you've experienced, please bring your awareness into the here and now and become fully present in this room, feeling very good!"

There will be a few minutes of bustle and hubbub when everyone stops Image Streaming. Wait for things to settle down, and then sound the chime several times, perhaps reminding participants, with a touch of humor, that the process is still under way and they still have perhaps the most important step in the experience left. Give everyone a 5-minute refreshment break; then have participants switch to fresh partners for the debriefing portion of the session.

Debriefing

1. "Any astronaut who returns from a mission to some far place has to be debriefed by the scientists back at Mission Control. The astronaut must relate every impression he or she can possibly remember from that experience so the scientists can discover more information than was originally noticed and reported when the events were happening. We can use this same method to extract more meaning from our Image Streams. In this session, think of yourself as an astronaut and your new partner as a scientist at Mission Control.

   "When you debrief to your partner, try to relate in just 2 minutes everything you observed in your own Image Streaming experience. Even though you are recounting from memory, tell it in the present tense as if it is happening right now. Instead of saying 'I *saw* such and such,' say things like 'I am now looking at such and such, which is colored with shadings of whatever, and its surface feels thus and so to me.' Try to report everything in such vivid, multisensory detail that your new partner won't be able to keep from sharing what you experienced. Only this time, in the debriefing, describe it with eyes open instead of closed. This will give you a new visual perspective on the experience, helping you build even more bridges between your left and right brains.

   "Debriefing is a very important step, because it is during this stage that many people have their 'Aha!' experience, when they suddenly realize the meaning or significance of what they saw and did in their Image Streams. You'll have 2 minutes to debrief, and then I'll sound the chime to signal turnaround and your partner will have 2 minutes to debrief to you in turn. Ready to debrief, set, go!"

Allow 3 minutes instead of 2 if this step is as animated as it usually is. Then sound the 30-second chime and finally the multiple chimes.

2. "With apologies for any incomplete debriefings, please now switch roles if you have not already done so, so your partner has time to debrief as well. Please resume with this debriefing now!"

Repeat the same process as before, ending with the 30-second warning chime and the final multiple chimes.

The Joking Analyst

"Some of you may have experienced an 'Aha!' insight during that debriefing, and others probably didn't. Whether you did or didn't, let us now do another technique called Joking Analyst, which has been shown to be very effective in extracting further meaning from the Image Streams.

"All the images you saw contain symbolic messages from your unconscious. Even if they are the most boring, ordinary-seeming images, there is a reason why your unconscious chose to show you those particular images and not others on this occasion.

"Humor is one of the best ways to unlock the hidden meaning of the images. I'm going to ask each of you to pretend that you are Carl Jung, Sigmund Freud, Milton Erickson, or some other great student of human symbols. In your role as a great psychiatrist, pretend that the Image Stream *you* have just experienced is a dream that a patient of yours has just related *to* you. Now pretend that your partner is one of your professional colleagues, with whom you are consulting. On my signal, I want you to begin discussing the so-called 'dream' with your partner, in the most pompous, ultraserious psychiatrist's manner you can imitate. Assume that every aspect of this Image Stream, no matter how banal, is charged with meaning. Speculate with your partner on the possible hidden meanings of your so-called patient's dream. Have fun with this. It is far more useful to have fun with this technique than it is to try to be right. Yet, paradoxically, the more fun you have and the sillier you get, the more likely you are to let down your guard and stumble unwittingly upon the true interpretation of your Image Stream. Begin now."

This should turn lively, especially in larger groups, where laughter and excitement tend to spread. Allow anywhere between 3 to 10 minutes, or until the buzzing and murmuring die down. Sound the 30-second warning chime and then the final multiple chimes.

Concluding Speech

"Thank you everyone. That was excellent. That concludes our Image Streaming for tonight. Image Streaming is a learned skill, like any other. As you continue to practice, you will find it easier and easier to see the images and to derive meaning from them.

"It's very important when you Image Stream to speak your descriptions aloud to an external focus. You don't always need a live partner. But if you practice Image Streaming on your own, you should speak your descriptions into a tape recorder. Just describe whatever ongoing Image Stream you see at the time onto blank tape, the way you were describing it here, to your partner. Always describe in the present tense and in rich, sensory detail.

"It says in this book that 10 to 15 minutes of Image Streaming per day will actually build your intelligence by reinforcing bridges of communication between the different poles of your brain. You can also learn to put questions to your Image Stream, and it will provide insightful answers in the form of imagery—answers far more profound than any you would get from your logical, one-word-at-a-time left brain.

"Please keep on Image Streaming, and please do come to our next session. Thank you for making this experience happen."

After you've closed the formal meeting, let people socialize for a while over refreshments, sharing observations and reactions. You can circulate among them, encouraging people to stop right in the middle of their conversation, close their eyes, and look in on their Image Stream to see what is playing there in this particular context. Such Instant Answer pictures may shed more light on the subject of conversation. The technique is also a fun way to end a session and gives people a deeper feel for the richness of the Image Streaming experience.

Do It Yourself!

No law says that you must use the script presented in this chapter, word for word. It is provided merely as a crutch and a guideline. The basic format can be easily adapted to every other technique described in this book, including Instant Replay, Beautiful-Scene Describe-Aloud, Borrowed Genius, and Parallel Earth. As you grow adept at the procedures and gain a feel for group dynamics, you will soon find it easy to write your own scripts, which will likely be as effective as mine—if not more so. (You can order premade audiotapes for many of my techniques from Project Renaissance, P.O. Box 332, Gaithersburg, MD 20884. Tel.: 301-948-1122 or 800-649-3800.)

THE BUZZ-GROUP PRINCIPLE

All of my group and partnering techniques derive from the Buzz-Group Principle. In the early days of brainstorming, executives would gather as many as thirty or forty people into a conference room and have them pitch ideas at random. They quickly learned that this method didn't work. Most people never got a chance to speak and were thus frustrated in their Socratic drive for self-expression.

Nowadays, brainstorming sessions are kept to a maximum of three to five people so everyone can have his say. When large groups are unavoidable, they are broken into small buzz groups, which pool their results later in the larger group.

Dynamic Format

My work with Project Renaissance has convinced me that the *process* of buzz-grouping is far more important than the results. Through Socratic self-expression, each participant arrives at insights that are profoundly important to him, whether or not they are shared with the group at large.

In my consultation work with corporations, I teach a version of buzz-grouping called Dynamic Format, in which the pooling of results from the buzz groups is entirely optional. The real benefits of Dynamic Format are felt in the weeks and months after the seminar as participants interact with one another more effectively day by day, each responding from his own individual experience of enlightenment.

Image Streaming workshops run on the same principle. The important work is done within buzz groups of two or three partners who alternately describe and discuss their respective Image Streams. It can sometimes be fun to bring everyone together at the end of the session so that different groups can find out what the others came up with. But there is little real gain from such workshopwide discussions, and always the danger lurks that some people may grow impatient and inattentive from too much passive listening. Keep such global sessions short, if you use them at all.

YOU ARE THE VANGUARD

Image Streaming was born from group interactions in the 1970s, when a handful of accelerative-learning enthusiasts gathered to experiment. It is still in group situations that Image Streaming reveals its most exciting and energizing insights.

Using techniques like High Thinktank and Borrowed Genius, group Image Streamers can brainstorm vital questions of the day, ranging from the national deficit to the AIDS education program at your local high school. I have found that when people come together in buzz groups to share their Image Streams on important issues, the resulting discussions are intense, pointed, and unusually productive. When necessary, your Image Streaming group can transform rapidly into a core of likeminded people dedicated to pursuing a specific goal, then shift just as rapidly to another focus once that goal is achieved. Whatever issue you tackle, your group will be linked by the sort of deep bonds that only form between people who meet together in the supercharged realm of the greater consciousness.

As Image Streaming and similar techniques become more widespread, we are fast becoming a nation of visionaries, all nurturing our private dreams of change. The time when we must tell our visions to the tribe is near. I urge you to reach out now and become the first in your community to launch an Image Streaming support group. Along with your friends, colleagues, and neighbors, you will form the vanguard of a truly democratic consensus, in which troublesome issues are addressed not by clashing factions spouting political slogans, but by free men and women of diverse opinions, joined by a spirit of empathy and guided by the wondrous insight of their own Image Streams.

CHAPTER 13

THE WINDOW OF CHILDHOOD

In 1995, University of Alabama neurologist Britt Anderson found that a section of Einstein's frontal cortex contained a higher density of neurons than did identical slices from five other brains.[1] Assuming that this higher density is consistent throughout Einstein's brain (Anderson does not assume this), does that mean that Einstein was born with more neurons than the rest of us?

Not necessarily. All of us are born with far more neurons than we need. Through the natural process of *apoptosis,* or "programmed cell death," massive numbers of our brain cells die off before the age of two. Einstein was likely born with a similar number of neurons to the rest of us. But, for some unknown reason, he may have lost fewer cells to apoptosis.

This is only one example of the kind of massive physical change that goes on constantly in children's brains. By regulating nutrition, stimulation, and other environmental factors, parents can intervene dramatically in their children's intellectual growth.

Researchers at Heinrich Heine University in Dusseldorf, for example, recently found that professional musicians who started music lessons before the age of seven had very different brains than those who started at age ten or later. Under magnetic resonance imaging, both groups were revealed to have a left *planum temporale*, a part of the temporal lobe that processes speech and other sounds, that was much bigger than their right *planum temporale*. This is a trait common to all people. But the group that started lessons younger showed a far greater degree of asymmetry than did the late starters.[2,3,4]

A WINDOW OF OPPORTUNITY

The Dusseldorf study, published in February 1995, is only the latest in a long series confirming that we can significantly modify the brains of children through early training and conditioning. It has long been known, for example, that children who are taught to play and sight-read music at an early age tend to be several standard deviations above average in intelligence.

We have already mentioned in previous chapters, as in the example of the Mankato nuns, the growing evidence that even the brains of the elderly can be sharply improved with proper conditioning. You are, in fact, never too old to increase your intellect. Still, there is little doubt that very young children are more susceptible than others to brain-enhancement techniques.

Peak brain growth occurs in three distinct spurts. The first begins 8 weeks after conception and continues until the thirteenth week. The second begins 10 weeks before birth and continues until age two. During this phase, most of the interconnections between neurons are formed. Finally, between the ages of two and five, the brain grows to 90 percent of its final weight.[5] The critical years between conception and age five therefore present an invaluable window of opportunity during which we can "jump-start" the brains of our children, giving them an unparalleled early advantage.

Sensitive Brains

Every parent knows that children are more sensitive than adults. They frighten more easily and hurt more sharply. At the same time, they also possess a keener feel for beauty and wonder. These qualities are linked to the child's uncommon learning potential. The same trait that makes children vulnerable to hurt also makes them open to beauty, learning, and knowledge.

Thin Boundaries

Some years ago, psychiatrist Ernest Hartmann grew curious as to why some people were more prone to scary dreams than others. Through his work as director of the Sleep Research Laboratory at Boston's Lemuel Shattuck Hospital, Hartmann knew that about 5 percent of the population was plagued by extremely vivid nightmares as often as once a week.

After studying about a hundred of these frequent nightmare victims, Hartmann discovered certain common traits. People who have unusually lifelike nightmares also report a similar vividness in their normal, nonfrightening dreams. They also remember their dreams more often than others do. All the test subjects reported having strong memories of early childhood, going back as far as their second and third years. They saw themselves as "unusual" and "different" and used modes of thinking that seemed to Hartmann "loose and tangential." Perhaps most important, a large proportion of Hartmann's test subjects worked in creative jobs as musicians, painters, writers, and the like.[6]

In his 1992 book *Boundaries in the Mind,* Hartmann concluded that people who were prone to nightmares had "thin boundaries"—an unusual sensitivity to people, forces, and stimuli outside themselves.

"I Can't Keep Things Out"

According to Hartmann, thin-boundaried people often find themselves swept away by the flood of sense and emotion that makes up normal life. Loud noises jar them. Bright lights dazzle them. Violent images scare them.

"I can't keep things out," complained one of Hartmann's test subjects, an artist in her late 20s.[7]

Hartmann found it "striking" that his thin-boundaried subjects "did not use typical, relatively mature defenses, such as isolation, intellectualization, repression. . . . I saw these subjects as 'defenseless,' or unusually vulnerable."[8] They exhibited, in short, an emotional sensitivity very much like that of children.

The Circuit Breaker

Sigmund Freud first proposed the idea of "ego boundaries"—psychic barriers that shield a person's fragile self-image from outside threats and repressed memories. But Hartmann suggests a deeper role for these "boundaries in the mind." He believes they may safeguard the orderly operation of the brain itself by locking different functions in their appropriate compartments, much as a dike holds back the sea.[9]

Without such boundaries, our brains might overload like a computer during a power surge. We can imagine these psychic boundaries as circuit breakers that cut off the juice when the current reaches a critical threshold.

Permeability

All of us had thinner boundaries when we were children, says Hartmann, though some more so than others. As we grow older, the need to protect ourselves from hurtful stimuli causes our boundaries to thicken. Most adults fall somewhere between the opposite poles of extreme thickness and thinness as measured by Hartmann's diagnostic Boundary Questionnaire. They are neither sensitive nor callous, neither wildly imaginative nor rigidly dogmatic.

Such a middling state is ideal for conforming to society. Unfortunately, it is less than ideal for ingenious thought and accelerated learning. Thin or permeable boundaries provide a key advantage in performing most of the techniques you have learned in this book. Thin-boundaried individuals, for example, frequently take on alien identities in their dreams, such as becoming animals or people of opposite sex, making them naturals for the Borrowed Genius technique. The vividness of their dreams suggests an equal ability to achieve powerful Image Streams.

In general, thin-boundaried folk seem more open to the unconscious, whence springs ingenious insight. The same circuit breakers that protect us from pain and confusion also tend to abort our most brilliant imaginings. Without extremely permeable boundaries, human genius simply cannot exist, for it arises from the commingling of the very sights, sounds, thoughts, and memories that our primitive defenses struggle so desperately to keep separate.

The trick lies in learning how to thicken or thin our boundaries at will in order to render them appropriate for changing circumstances. The thickening part is easy—all too easy, in fact.

THE THICKENING PROCESS

Much of the training to which we subject small children has the effect of thickening their boundaries. In one of our recent *Capital Ideasmiths* newsletters, my wife, Susan, enumerated some of the many ways that parents and teachers thicken children's boundaries without even realizing it. How many times, for example, have we told our children, "Don't play with your food!"

"Food is the most sensuous experience children have," Susan writes. "It arouses all five senses. How can any intelligent, feeling, inquisitive, experiential person not play with his jello? But no one wants to see food played with. And, it is convenient for Mom to clear all the dishes off the table at the same time. So, 'Don't play with your food!' "

Then there is the all-important training to sit up, fix your eyes on the teacher, and "pay attention" in school.

"You can't absorb new ideas without thinking about them. If your eyeballs don't roll up into your head, if you don't gaze out the window, if you don't doodle, you are not learning ideas, you're memorizing facts. . . .

"As children get older, they internalize these instructions. They teach themselves to ignore the images that the brain by its nature forms, to stand up straight, to focus their attention, to not play with their food, to not daydream."

CONTROLLED FREEDOM

Obviously, we cannot allow children to make messes at will or to ignore the teacher in class. Such license will degenerate into appalling and dangerous indiscipline as the child grows older. Somehow, we must impart to children the discipline of civilized and orderly learning without thickening their boundaries to the point of stunting the imagination.

We can do this through a kind of educational judo. Recall from Chapter 10 our discussion of how to redirect people's natural self-expressiveness into a highly effective form of listening to others called Freenoting. We can work a similar magic with children, diverting their free-roving curiosity into methods of exploration that neither make messes nor disrupt classrooms.

THE MONTESSORI METHOD

At the turn of the century, Maria Montessori, an Italian physician, began working with retarded children at the University of Rome's psychiatric clinic. She quickly noticed that these "unteachable" children had an amazing capacity for concentration when they were focused on some play or activity that caught their interest. Only when a teacher tried to force their attention to less appealing subjects did they balk. Parents who have watched their children practice video games for hours on end understand this principle perfectly.

Montessori developed a new kind of classroom in which children could roam at will from one activity to another, focusing on whatever caught their attention most. She equipped the classroom with interactive learning aids that taught through all five senses. Children learned the alphabet, for example, not only by sight, but also by feel, identifying sandpaper letters with their eyes closed. They put together wooden maps to learn geography and explored math concepts by assembling rods, blocks, and other shapes with varying textures. The teacher's job was to circulate around the classroom, keeping an eye on things and making sure the children followed the rules—such as keeping their voices down so they wouldn't disturb the other children's "work."

Although the Montessori Method was first used successfully on retarded children and poor children in the slums of Rome, most people today think of it as a costly luxury for affluent families. This is unfortunate, for I believe the method would serve our own urban poor quite well.

Today the Montessori Method is widely used in many countries, but it came late to the United States, with the first authorized Montessori public school opening in Cincinnati, Ohio, only in 1975. There are now about 3,500 Montessori preschools in the United States, and about 100 U.S. elementary schools use the method.

Visitors to Montessori schools are often struck by the silence. Even preschool children remain quietly absorbed in their tasks for hours at a time. Their concentration pays off. Montessori children typically learn to read and write well before the age of five.

The Triumph of "Teaching"

The secret of the Montessori Method is the Feedback Principle. Montessori children receive constant feedback from their acts of spontaneous self-expression (the key component of the Expression Circuit, to which we traced much of the brain's development, both physical and mental, in Chapter 4).[10] University of Pittsburgh sociologist Omar K. Moore achieved similar results by designing classrooms that interacted with children through such devices as the Edison Talking Typewriter. Moore easily trained two- and three-year-olds to read, write, and even type.[11] Yet, like the Montessori Method, his interactive technique failed to take hold in a society that had long ago abandoned Socratic Method in favor of didactic teaching.

Games and Order

Anyone who has ever played a game with small children has noted how vigilant they are for even the slightest infraction of the rules. Children have a deep love for order. They sense instinctively that "a game without rules is no game at all," as David Kahn, executive director of the North American Montessori Teachers' Association in Cleveland, puts it.[12] Although Montessori children are free to roam the classroom, choosing tasks at will, they adhere to a strict protocol of courteous behavior and progress through a highly structured curriculum toward well-defined goals. They do not make messes, nor do they disrespect their teachers.

Not everyone has access to a Montessori school, nor does this method solve every ill of our educational system. Yet it has succeeded brilliantly in balancing freedom with discipline, allowing children free rein to explore their world without trampling on others' rights. This sort of educational judo forms the core of any successful program of accelerated learning for children.

TEACHING YOUR CHILD TO IMAGE STREAM

One of the most educational "games" you can possibly teach your child is the technique of Image Streaming. It offers children unlimited freedom to roam through their imaginations, while at the same time building their intelligence and reinforcing the thin-boundaried behavior of imaginative thinking. Unlike playing with jello (or decorating the room with crayons), Image Streaming makes no messes. Moreover, it disciplines the child to set aside specific times for imaginative thought, leaving other times open for listening politely to the teacher.

The Game of Image Streaming

Even newborn infants have the capacity to form mental imagery, as evidenced by their eye movements during the REM phase of sleep. As soon as children are able to understand simple instructions, at about age three, they are ready to play the game of Image Streaming.

You should begin by demonstrating the technique to the child. A good approach is to say something like "You know, I think that we're dreaming all the time, not just when we are asleep. For example, even now, if I simply close my eyes, I see. . . ." Here you actually close your eyes and describe whatever image you see. Show the child exactly what is expected of him by describing your imagery richly and at length.

"I see the trunk of an enormous tree," you might say. "I think it is pine. It has that kind of close-knit bark, almost blond bark in color. The tree must be four feet thick. The ground all around has a thick cushion of pine needles. I see it continuing under other pines farther away, getting darker as it leads deeper into the forest. Behind me is a stream, just a really little stream, just a few inches across, and I can see that the ground under the pine needles is a light clay."

At this point, the child will be fascinated by your seemingly magical ability to "dream" while awake. That's when you ask, "When you close your eyes, what do *you* see?"

Each time I have taught young children to Image Stream, I have been awed by the ease with which they see and describe images. Very young children have not yet been taught to squelch their mental imagery. Entirely free of the thick boundaries that block so many adults, children instinctively grasp the Image Streaming method and easily duplicate the feat.

Very young children may, at first, simply name the object or person they are seeing in their mind's eye. You may have to coax the descriptive details out of them through patient questioning. You can help your child build descriptive skills by playing a game of pointing to objects and asking the child to describe them or by having your child describe an object to you—but not name it—with the goal of getting you to guess which object in the room or yard the child is describing.

Unlike adults, children will need no excuse to Image Stream. You won't need to explain to them that its unique Pole-Bridging effects will help them excel in school. Once they discover the magic and fun of dreaming while awake, they will become as absorbed in this quiet, beneficial discipline as other children do in video games. Unlike television or video games, however, Image Streaming gives them the great enjoyment of being the focus of your attention during the game. By continuing this game throughout their childhood, you will greatly lessen the chance of developing a communications gap when your children become teenagers.

Every child I have ever taught to Image Stream has made quick and obvious leaps in perceptiveness and understanding. Few pleasures surpass that of seeing a child experience such a jump start in intellectual development, and knowing that you helped bring it about.

MUSICAL POLE-BRIDGING

One of the easiest ways to open your child's Image Stream is through music. Playing and listening to music are not only powerful Pole-Bridging techniques in their own right, but in experiments have been shown to stimulate the brain's image-generating faculties.

Researchers at the Montreal Neurological Institute, for example, subjected twelve people to various melodic patterns, while scanning their brains with positron emission tomography (PET), which tracks blood flow. When the music started, blood rushed not only to the right temporal lobe involved in hearing, but also to that region at the back of the right hemisphere that governs vision. Since subjects kept their eyes closed during the experiment, researchers concluded that the music must have automatically stimulated mental images.[13]

In Chapter 2, we discussed how synthesthetes see processions of pleasing, abstract shapes and forms when they hear music. The Montreal experiment suggests that all of us may react synesthetically to music, though, in most cases, we squelch the images before they reach our conscious minds.

Einstein's Violin

Scientists have found that certain types of music act as powerful stimulants to intellectual development, for children and adults alike. As the Dusseldorf experiment showed, however, it is during childhood that the impact of music is most acute.

Albert Einstein was a passionate violinist for most of his life. Of his music and physics research, Einstein said, "Both are born of the same source and complement each other . . ." Einstein's relatives observed that music seemed to catalyze his creative process.

"Whenever he felt that he had come to the end of the road or into a difficult situation in his work," his oldest son remembered, "he would take refuge in music, and that would usually resolve all his difficulties."[14]

Einstein's sister remarked that playing music seemed to "put him in a peaceful state of mind, which facilitated his reflection." While puzzling over a physics problem, Einstein would play his violin until, suddenly, he would stand up and declare, "There, now I've got it!"

"A solution had suddenly appeared to him," his sister observed.[15]

Playing music calmed Einstein's soul and opened pathways to his subconscious. But it may have done more. Einstein took up the violin at the age of six, when his window of childhood was still wide open.[16] By the time he was fourteen, he was playing Beethoven and Mozart sonatas, as well as spending many hours improvising on the piano.[17] Einstein may have unwittingly exposed himself to one of the strongest intelligence builders of all, during the very years of his childhood when it could do the most good. A large portion of Einstein's prodigious intellect as an adult may have resulted from this timely childhood training.

MOZART RAISES IQ

Scientists at the Center for Neurobiology of Learning and Memory at the University of California at Irvine subjected thirty-six college students to a battery of spatial IQ tests—that portion of the standard IQ test that measures right-brain ability. After taking the test, students listened for 10 minutes to Mozart's *Sonata for Two Pianos in D Major,* K. 448. Immediately after, they were retested and were found to have IQ scores 8 or 9 points higher than before.[18,19,20] This IQ gain faded after about 15 minutes, but the researchers suggested that actually *playing* music might create a more prolonged improvement.[21] Other researchers suggest that regular listening may also prolong the IQ gain.

These researchers did not use the word *Pole-Bridging* in their attempts to explain the phenomenon, but one of them, Gordon Shaw, did guess that the complexity of the music somehow primed areas of the brain involved in abstract reasoning and other nonmusical tasks.[22]

The Lozanov Breakthrough

While completing his doctorate at Kharkov University in the Soviet Ukraine, Bulgarian psychologist Georgi Lozanov made a remarkable discovery. He had been studying the various methods used in the Soviet Bloc for accelerated learning, including hypnosis, sleep learning, and even yoga. These techniques had proved effective but were difficult and cumbersome to use.

Then Lozanov learned that many Soviet-Bloc hospitals were using music to soothe and relax patients. Back home in Bulgaria, Lozanov and a colleague, Dr. Aleko Novakov, began testing the effects of various types of music on the learning process. U.S. researchers Lynn Cooper and Milton Erickson had already shown that listening to metronomes set at 60 beats per minute would induce in listeners a profound Alpha State, ideal for learning and memory. Apparently unaware of this American research, Lozanov and Novakov discovered independently that slow Baroque music, with 60 to 64 beats per minute, would also induce an Alpha State. Baroque music was popular in Europe between 1600 and 1750, ending with the death of its most celebrated practitioner, Johann Sebastian Bach. Baroque music is characterized more by rich texture and consistent rhythm than by melody or contrasting tempos.

Lozanov and Novakov discovered that people subjected to Baroque music absorbed information as efficiently as sleep-learners, a breakthrough that quickly led to the development of Lozanov's famous Suggestopedia method, by which information—such as phrases in a foreign language—is fed at 4-second intervals to the learner, against a background of 60-beat-per-minute Baroque music. Early results showed that Suggestopedic students could absorb anywhere from 60 to 500 words of a foreign language per day. The Soviet-Bloc press was soon boasting that Suggestopedic learners could routinely master foreign languages in a single month.[23,24]

Many U.S. researchers dismissed these claims as Communist propaganda. However, researchers at Iowa State University soon succeeded in increasing memory retention in test subjects by 26 percent and speed of learning by 24 percent using 60-beat-per-minute Baroque music.[25]

Popularized in the United States by the 1979 best-seller *Superlearning,* by Sheila Ostrander and Lynn Schroeder, Suggestopedic courses are offered today by licensees throughout the United States. Several vendors market tapes of Baroque music specially chosen to conform to Lozanov's standards. A reliable supplier is Superlearning, Inc., in New York City (tel: 212-279-8450).

The Mozart Connection

The ear plays a key role in electrically charging the brain, according to Dr. Alfred Tomatis, a respected member of the French Academy of Medicine and Academy of Science. When the electrical potential of brain cells starts to fade, says Tomatis, we experience dullness and fatigue. Like batteries, brain cells must be recharged. Tomatis discovered that one way to accomplish this is through listening to high-frequency sounds, between 5,000 and 8,000 hertz. According to Tomatis, the vibration of Corti cells, which line the fluid-filled cochleas of the ear, acts as a kind of brain generator.

After years of analysis, Tomatis concluded that the music of Mozart contained the highest number of sounds in this frequency range, while hard rock contained the fewest. He also recommends Baroque music and Gregorian chant for the purpose of recharging the brain.[26]

While useful for people of all ages, the discoveries of Lozanov, Novakov, and Tomatis have particular value for enhancing the brainpower of children, both before and after birth. Their theories have given rise to a wide range of simple but effective interventions, several of which are included in the following pages.

PRENATAL STIMULATION

The months before birth are the time when you can make the greatest difference in your child's future. Ten weeks before birth, the brain is already forming interconnections between neurons that will determine the baby's intelligence and manner of thinking. This is your optimum window of opportunity to enhance your child's intellect.

Mozart in the Womb

It is widely accepted today that unborn babies can hear much of what goes on around them in the room, including conversations. However, high-frequency sounds normally are filtered out by the skin and the uterine wall. Unfortunately, it is these high-frequency sounds—the ones that Tomatis credits with recharging the brain—that carry the most information, in speech as well as in music.

You can overcome this barrier by placing stereo headphones on the womb area. The headphones will convey the full range of frequencies through the body, allowing you to pipe in everything from Mozart to your own voice (provided your sound system will allow you to broadcast through a microphone).

Don't blast the sound. Normal volume, such as you would use to listen to music through the headphones, will do. Remember, the fetus has no way of escaping or blocking its ears if the sound is too loud for comfort.

Mother-Baby Feedback Loops

You don't have to wait for your baby to be born before you can start engaging it in a stimulating feedback loop of self-expression and response.

One way to interact with your unborn child is through light. If you bare your abdomen to sunshine or some other very strong light, you can actually catch the fetus's attention by playing shadow games with your hands. Don't do this for too long, though, because the light may prove overwhelming to the child.

You can also catch the child's attention by gently massaging the womb area. When the baby kicks, make gentle taps in response. Tapping on the sides of the tub while bathing works even better, because the sound is carried better through liquid.

Through persistent practice of these techniques, you will be amazed and delighted at the degree of interaction you can achieve with your unborn baby, stimulating a precocious awareness of the outside world.

Increased Blood Flow

Fetuses are just as dependent as adults on blood flow for their brain development. Some years ago, a group of South African physicians marketed a birth suit, which they claimed could elevate the potential IQ of the unborn child by as much as fifteen to twenty points. The birth suit enclosed the lower abdomen in an airtight chamber, subjecting the womb area to reduced air pressure for short intervals, thus causing extra blood to circulate in the area. This intriguing device is no longer on the market. However, Project Renaissance has developed a psychegenic method (the opposite of psych*o*genic, which refers to thought processes that lead to bodily disorders) that promotes a similar effect.

Researchers in the field of biofeedback have ascertained that virtually any part of the body on which you focus attention will receive a somewhat increased blood flow. You can amplify this effect through the following visualization.

Decide which part of your body you wish to enrich with greater blood flow. Now imagine that part of your body to be an inch larger than it is and an inch farther out from the center of the body. Get in touch with the physical sensation of how it would feel if that part of the body were actually an inch larger than usual. This exercise will distort the brain's body image, tricking the brain into rushing extra circulation to the supposedly larger area. Trying to increase the size of most body parts by more than 2 inches will not work, perhaps because the brain recognizes that extension as unrealistic. However, you can imagine the hands and feet as far larger than an inch over actual size and receive a correspondingly large enrichment of blood, perhaps because we are used to reaching out and interacting with the physical world through our hands and feet and feel that we have more objective control over them.

Belief in the technique is, of course, helpful in making it work. If you are skeptical, try walking outside in cold weather and using the method on some extremity of your body. The uncanny warming that occurs in the target area will do wonders for dispelling your doubts on this point.

A pregnant woman should imagine her womb and fetus to be enlarged for 2 to 3 minutes at a time several times a day. This simple exercise will supercharge the baby's growing brain with oxygen at the very time it will do the most good. Just as important, the increased blood flow will help drain toxins that could easily damage the baby's budding nervous system.

Avoiding Toxins

Advice on nutrition during pregnancy has become something of a cottage industry. Not being an expert in this field, I will refrain from adding to the wealth of information already available. I do, however, recommend special attention to prenatal nutrition. Two sources of information are Adele Davis's *Let's Have Healthy Children* (New York: Harcourt, Brace Jovanovich) and Richard Passwater's *Supernutrition* (New York: Dial Press).

I would be remiss if I failed to touch on the issue of toxins, however briefly. Toxins can present a major threat to your baby's nervous system. Fetuses are notoriously helpless against poisonous substances of all kinds. They have virtually no capability of cleansing their systems, once invaded. For that reason, the mother must avoid all food, drink, and other substances that tend to pollute the bloodstream.

Forget about weight reduction during pregnancy. When the body breaks down fat, toxins (such as insecticides) that have been stored in the fat for years will dissolve into the bloodstream and go straight to your baby.

Drugs, alcohol, tobacco, and coffee are also out of the question. Many doctors will tell you that a drink now and then won't hurt. What they mean is that it will not seriously harm your child. However, it will very likely impact his future intelligence. Is an occasional drink really worth it?

Avoid any substance to which you may have allergies, especially food dyes and additives. Stick to whole grains and well-washed raw fruits and vegetables as much as possible. The best time to visit an allergist is before you become pregnant, so you'll know exactly which substances to avoid.

Obviously, these guidelines are all subject to your doctor's advice. For medical reasons, you may have to eat certain foods, stick to a certain weight, or take certain drugs that would not necessarily be optimum for your baby's brain growth. The best insurance is prevention. Being in the best possible health *before* you get pregnant will minimize the need to put your baby's future IQ at risk.

INFANT STIMULATION

It is during infancy—from birth to age two—that most of the dendrites and axons connecting your baby's neurons to each other are formed. Here are a few techniques that will help you accelerate and enrich this process.

The Babinsky Reflex

Every baby is born with the Babinsky Reflex, an instinctive urge to crawl on all fours when the bottom of the foot is tickled. Millions of years ago, when babies were placed stomach-down in ticklish beds of leaves or grass, the Babinsky Reflex got them moving as soon as possible, much as the feel of seawater inspires newborn whales and dolphins to swim to the surface for their first breath of air.

As we discussed in Chapter 4, crawling is one of the chief ways by which a baby gains feedback from the environment. Cultures that swaddle infants heavily or otherwise confine them are unknowingly hampering the intellectual growth of their newborns.

Place your baby on his stomach and tickle the bottom of each foot in turn. This will elicit faint crawling motions at first. Later, it will stimulate your baby to actually start crawling.

"Find-Me" Mitts

The sooner your baby discovers his hands, the sooner he will gain the benefit of visual feedback from watching those hands crawl across the floor and manipulate objects. You can help by providing brightly colored "find-me mitts." These were invented by Dr. Burton White, whose company, Playtentials, unfortunately went out of business.

You can make your own find-me mitts by cutting a pair of baby mittens so the fingers are free and the mitt fits only around the palm. Color the mitts a very bright red, or decorate them with red and white stripes. You can also make find-me booties for the feet. Use these devices only until your baby discovers his extremities, and then stop using them. They tend to reduce the important tactile input into the hands and feet when the baby is crawling.

It is an established practice in many hospital nurseries to place a red balloon over a baby's crib so he can see it.

From day one, newborn babies are able to see. The balloon provides visual context in the starkness of the hospital nursery. Once home, you can do much better. Tie a red helium balloon loosely to your baby's wrist for 2 minutes, once or twice a day, so that his own random movements will effect an interesting change in his visual environment. Soon he begins to notice the connection. He now moves that hand more than the other and no longer at random, because he enjoys the feedback. Switch back and forth, from one hand to the other, to balance development.

In general, frequent massage, stroking, and gentle physical play of all kinds enhance your baby's opportunity for physical feedback and spur mental development.

Creating a Musical Environment

You should continue to expose your child to brain-activating music long after he is born. An environment rich in Baroque music, Gregorian chant, and Mozart will benefit a child of any age. I don't know whether Drs. Lozanov and Tomatis would agree, but my experience suggests that virtually any complexly layered and nuanced music will stimulate the mind, from Bach's Fifth Brandenburg Concerto or Schubert's Trout Quintet to progressive jazz.

Needless to say, serenading your baby yourself with a musical instrument, as well as the time-honored practice of singing lullabies and nursery songs, will be of great benefit.

Underwater Swimming

For all the reasons discussed in Chapter 11, underwater swimming is just as stimulating to the brains of children as it is to those of adults. It's never too early to start. Dr. Frederick LeBoyer, for example, advocates underwater immersion of babies from birth![27] A skilled swimming instructor can teach your baby to swim in the first days, weeks, or months. Play underwater fetching games with your toddler. From the time your child is old enough to understand, emphasize to him the benefits of swimming underwater and increasing the time during which he can hold his breath. I recommend Bonnie Pruden's program of physical activities for children, which includes such early swimming.[28]

EARLY CHILDHOOD STIMULATION

As previously discussed, it is during the early childhood years from 2 to 5 that your child's brain grows up to 90 percent of its adult weight. It is also during this period that your child's IQ will solidify at a particular level. Below are a few proven techniques for jumpstarting your child's intellectual growth during these critical years.

Early Reading

You can and should begin teaching your child to read before he is two, at the same time he is learning to talk. As with speech, he learns whole-pattern recognition, rather than breaking down language into its components. He learns to talk by assimilating whole phrases at a time. Similarly, he will most easily grasp written language by gulping down whole words at once. Learning individual letters or phonics will not help a one- or two-year-old. Flash cards containing whole words are the best approach at this age.

There are many methods for teaching very young children reading and other precocious skills. My favorite is described in Dr. Glenn Doman's *How to Teach Your Baby to Read* (Random House, New York) (available from the bookstores of the Institutes for the Achievement of Human Potential, 8801 Stenton Avenue, Philadelphia, PA 19118). Another good Doman book is *How to Teach Your Baby Math* (Avery Publishing Group, Garden City Park, NY, 4th Edition, 1994).

Sight-Reading Music

I strongly recommend that children be trained in sight-reading at the earliest possible age. As noted previously, early sight-readers have been shown consistently to test well above average in intelligence.

The Pole-Bridging effect of music is increased when predominantly right-brain musical skills are merged with left-brain skills such as the learning of musical theory and notation. The professional musicians whose brains were scanned in Dusseldorf were found to have the greatest asymmetry on the left side of their brains, not the right—a product, no doubt, of their years of formal training. Even passive listening to music is bound to have a more pronounced effect on intelligence if the listener has acquired a formal or left-brain appreciation of its nuances.

The Piano Game

Some years ago, my wife, Susan, invented an effective method for teaching sight-reading to children from one to five years of age.[29] Perfect pitch—the ability to play or sing a particular note, such as "middle 'C'," without hearing a reference note first—is considered a rare gift, even among musicians. Yet, if you start the Piano Game early enough, you will increase your child's chances of developing perfect pitch. The game, whose directions follow, should be played for 2 to 5 minutes per day.

1. Face the child away from the piano or other keyboard instrument.
2. Sound a single note on the piano while singing the name of the note, A, B, or whatever. Use natural notes—the white keys—at first. Flats and sharps can wait until later in the training. (You should, however, name every note that the child hits by accident on a miss, even if it is a sharp or a flat.)
3. After hearing you play the note, the child should run to the keyboard and try to hit the same note, if possible, on

the first try. When he strikes an incorrect note, say or sing the name of the target note. This will be a sufficient indication that the child has missed. "Hits," however, should be strongly reinforced with laughter, applause, and hugs.

4. Before each new round of the game, set a large 3-by-6-inch card on the music rack above the keyboard, bearing a short segment of base and treble clef bars. On these bars, the note you're about to hit should appear prominently.

   Don't call attention to the card. Simply change it each time to the next note you're about to hit. It may take several weeks for your child to catch on to the fact that the card has something to do with the note you are hitting. Only when he asks about it do you briefly explain that the note on the card indicates the note you are playing on the keyboard.

5. Once the child has mastered sharps and flats, you can move on to sequences of two and three notes. This will soon lead to the child's being able to make out simple tunes and chords from musical notation.

A child trained in this manner will be well-prepared for conventional music lessons later on. However, this is not the main purpose of the exercise. Whether or not you want your child to become a musician, training him in perfect pitch and sight-reading skills will build his intelligence, possibly permanently raising the child's IQ more than ten points according to some evidence. Note that the *planum temporale*, that part of the brain that grows larger with perfect pitch, also governs word comprehension and certain mathematical skills.

The Game of "Velvety Smooth"

Psychologists have long known that seemingly slight traumas experienced in the first two years of life can haunt a child for decades to come. Most adults are plagued by a poor image of themselves and of their bodies, dating back to their earliest years. You can do much to erase such troublesome obsessions from your child's mind, freeing him to excel and reach his potential. An excellent method for doing this is the game of Velvety-Smooth.

Find a garment or a piece of rich, deep velvet or silk, preferably of an attractive color. Encourage the child to feel it, admire it, and enjoy its rich smoothness. Then say to the child, "Now I'm going to make my own breathing feel that way to me, all velvety smooth and just like these slow, smooth, rich strokes. Like this. . . ."

Let the child watch you as you spend a minute or so taking in a deep, smooth breath and obviously enjoying it to the hilt. "Um-m-m-m-m," you might say. "That makes me feel velvety smooth all over. Now, Johnny, can you breathe all velvety smooth like this?"

Your eventual goal is to get the child to feel his entire body as "velvety smooth," feeling good all over. At the same session, or perhaps at a later one, you can then ask, "Now, Johnny, is there any part of your body that doesn't yet feel velvety smooth?"

Maybe there won't be. But if the child indicates a part of his body, encourage him to talk about how it feels, to describe and put a name on the feeling if he can. With luck, the child may even bring to mind some past or present situation that caused hurt or difficulty in connection with that part of the body. Don't put words in his mouth. Make no suggestion of any hurt or difficulty or anything negative. Children are extremely suggestible, and your expectation of negativity, even if unspoken, could lead the child to construct some problem that didn't exist before.

If the child does identify some troublesome experience and the two of you have talked it out, encourage the child to resume his breathing exercise and to make that part of the body and that troublesome memory "all velvety smooth."

If all goes well and the child's interest is holding up, you can probe further, asking him to remember an even earlier experience that somehow felt the same way. You will sometimes uncover surprisingly early experiences! Never press such matters. Proceed in this direction only if the child seems ready and interested.

One good way to enhance the velvety-smooth experience is to take some of the more pleasant-smelling spices from your spice rack, such as cinnamon, vanilla, or peppermint. After the child has worked through whatever problem was troubling him, allow him to smell the spices while he returns to the velvety-smooth state. This added sensory dimension will further anchor his feeling of well-being. If this spice-smelling exercise is done in front of a mirror, your child will associate the happy ending with his own image. Every time he looks in the mirror as an adult, his own reflection will trigger some vestige of the feeling of peace, balance, and well-being you are giving him now.

INSTILLING CONFIDENCE

Any expert who presumes to tell parents how to raise their children will inevitably find his own offspring scrutinized for evidence that his techniques really work. I wouldn't presume to offer my family as an example of perfection, yet I can say that my daughter is a MENSA member, having been admitted to MENSA at the age of three, and was for a time the youngest member in that organization's history. (MENSA admits only people with IQ scores in the "genius" range.)

As a parent, I have found that no technique can, in itself, make your child a genius. Your *unspoken* feelings, beliefs, and attitudes will most impact your child's self-image and hence his potential.

Few parents need to be told to love and believe in their children. Our offspring hold an instinctive fascination for us. We cannot help marveling at their slightest achievements, especially in the first few weeks. Yet, as the demands of feeding, training, and disciplining the child grow, it is all too easy to forget that initial wonder and allow negative feelings to creep in. Without realizing it, we program a troublesome child into feeling inadequate or unloved—feelings that, once instilled, will return again and again in later life to steal his confidence or block his learning.

As I hope I have made clear throughout this book, we humans are marvelous creations, filled with awesome potential. Only negative beliefs hold us back, obstacles that have so far not been instilled in your infant. Remember each day to spend time simply gazing at your child and pondering the incredible mystery of his destiny and the boundlessness of his capability. Take delight in the sheer magic of childhood. Imagine what it must be like to see life through those fresh young eyes. Your sense of wonder will communicate itself to your child in ways more subtle and mysterious than we can fathom, and your child will grow to fill those wondrous shoes you have cobbled for him.

CHAPTER 14

USE THE FORCE

In the climactic moment in the movie *Star Wars,* Luke Skywalker swoops his spacecraft down into the intricate labyrinth of trenches and passageways that honeycomb the evil Death Star. Luke is faced with an impossible task. Hurtling at blinding speed through the narrow passages, he must drop a bomb down a tiny aperture that leads directly into the Death Star's core. His aim must be perfect. He will have only one shot at the target. As he approaches the critical moment, Luke fiddles frantically with his scanners, bombsights, and other high-tech instruments.

Then suddenly, out of nowhere, comes the ghostly voice of Luke's old mentor, the Jedi knight Obi-wan Kenobi.

"Use The Force," says Obi-wan.

Luke understands. To the distress of his squadron commander, Luke suddenly shuts down all his technical tracking equipment. He relaxes, lets go, and allows The Force to take over. His bomb falls right on target. The Death Star is destroyed, and the galaxy is saved from tyranny.

"DOES IT FEEL RIGHT?"

The Force is far more than a figment of filmmaker George Lucas' imagination. It is a real factor in human achievement. Military behavioral experts even have a name for Luke Skywalker's daring tactic. They would say that Luke used the "K check" to drop his bomb.

The Stinger missile is one of the most sophisticated "smart" weapons in today's high-tech arsenal. Fired from the shoulder, it beams in on its target by an infrared scanner. Once the Stinger is locked on, it can outmaneuver a jet fighter.

Yet, for all its bits, bytes, and microchips, the Stinger depends upon the intuition of its human operators. Expert Stinger shooters report that, just after hearing the beep that means they have acquired the target and just before pulling the trigger, they always stop and ask themselves, "Does it feel right?" They have learned in the field that if you fire the missile when it feels wrong, you miss. But when it feels right, you hit your mark.

Military researchers named this procedure the K check, or kinesthetic check. Nobody really understands how it works. In some extraordinary way, the eye, the mind, and the body must cooperate unconsciously to determine the most accurate trajectory for the missile. Finding the right trajectory involves taking into account the speed, size, and range of the target, the speed of the missile, the timing and angle of its firing, and even the anticipated action of its homing technology. Any *conscious* attempt to compute this many variables would overwhelm even an Einstein. Yet ordinary soldiers—including the illiterate Mujaheddin partisans who used the Stinger to sweep the Afghan skies of Russian helicopters—perform the job easily and consistently under combat conditions. Such is the power of The Force.[1,2]

What is The Force? The easy answer is that The Force is the power of our subconscious minds. After all, no conscious calculation could duplicate the K check. As obvious as this might seem, however, it is only half the answer.

Neither Luke Skywalker nor the Afghan Mujaheddin would have hit their targets had they relied upon intuition alone. Both had to spend long hours mastering their weapons before they could afford the luxury of letting go. Most of us have had the experience of daydreaming at the wheel of our car only to realize, with a start, that we have suddenly arrived at our destination. Such absentmindedness would be a serious threat to life and limb had we not previously mastered the skill of driving. At such times, it is The Force that brings us safely to our destination.

The Force is nothing less than that majestic freedom that comes to us only when we have mastered basic disciplines and technical skills. It is the power unleashed when right and left hemispheres work together in perfect harmony.

IN THE BEGINNING WAS THE WORD

According to the Bible, the universe came into being only when God uttered the appropriate words for "earth," "ocean," "light," "heaven," and so on. The ancient Egyptians believed similarly that their god Ptah had literally spoken the universe into existence through the utterance of *hekau*, or "words of power."

"Indeed," states a 4,000-year-old Egyptian text, "every word of the god really came into being through what the heart thought and the tongue commanded."[3]

These creation stories provide an apt metaphor for the power of our left brains, which think predominantly in words. In our mind, as well as in God's, the articulation of a thing in words will always make it more real, lending clarity, order, and substance to what would otherwise appear vaporous and ethereal.

Words and Pictures

In a famous experiment, the Russian psycholinguist Lev Vygotsky instructed very young children to draw butterfly wings. Those children whose budding vocabularies already contained words for "dot," "triangle," "slash," and other basic forms were able to sketch the wings quite well, even from memory. Those who lacked knowledge of these words could not draw a convincing pair of wings even when they were copying from a picture.

Vygotsky then took half of those children who had failed to draw the wings and taught them the crucial words. The other half of the group was allowed to remain ignorant of those words. When Vygotsky repeated the experiment, the newly enlightened group was able to draw butterfly wings just as well as the previously successful group. Those children who remained ignorant of the words showed little improvement from the first round.[4]

We see a similar effect in Image Streaming when our verbal descriptions actually cause the mental images to become more vivid and tangible. In a very real sense, our words bring the universe into being—at least within the boundaries of our minds.

THE BIPEDAL MIND

Until now, we have tended to view the left brain almost as an adversary, as a Squelcher to be evaded and outwitted. Yet, a genius who cannot put his discoveries into words (or, by extension, into mathematics or other forms of orderly expression) cannot truly create. His genius is useless to others, for it cannot be transmitted.

We might think of the brain as walking on two feet, like a man. The creative right brain and the judgmental left brain each represent one of those feet. Unless both feet move, we stand little chance of getting anywhere, yet both cannot move at once. One must stand still while the other swings forward, each in its proper turn.

Alex Osborn, the inventor of brainstorming, expressed the same idea in terms of driving a car. We won't get very far if our only strategy for driving is to press down on the brake, said Osborn. Yet imagine the trouble we'd be in if the only pedal we knew how to use was the accelerator! In order to get anywhere, we must alternate constantly between creation and analysis, between right and left brain.

Break Past the Bottleneck

Many readers, when they started this book, were like a man who walks with small, cautious steps. If you have been practicing your Image Streaming diligently, you can, by now, take much larger steps—but only with your right leg! Until you build up both legs equally, you will still have to take small steps.

The eighteenth-century economist David Ricardo expressed this principle in his famous Law of Variable Proportions, popularly known as the law of diminishing returns. This law states that the more you push to increase production, the smaller your production gains will be. This happens because our efforts at improvement are usually lopsided. We might double the number of workers in a factory but fail to replace the one inefficient assembly machine through which all the parts have to pass. Unassembled parts then pile up to the ceiling in front of that machine. You can't break through this bottleneck by hiring more workers. You have to buy more machines.

This, then, is the Bottleneck Principle, a corollary of Ricardo's law: When improvements in one sector create a bottleneck, the only way to break through that bottleneck is to make corresponding improvements in some other sector. Likewise, Image Streaming and the right-brain skills that it strengthens will take you only as far as your first big bottleneck. To break through to high-level genius, you must make corresponding improvements in your left-brain skills.

READIN', 'RITIN', AND 'RITHMETIC

Unfortunately, most of us learned in school to squelch our love for words and numbers. As a result, few American adults have a satisfactory grasp of the three Rs. Image Streaming alone will not let us break through that bottleneck. We must involve the left brain.

Maria Montessori, as we noted in Chapter 13, observed that children approach their play with a seriousness and mature concentration that would be the envy of any scientist or scholar. Those children—like young Albert Einstein—who go on to become consummate masters of the three Rs are invariably those same lucky few who learned early to regard those left-brain skills as a form of play.

A Merry Science

Einstein had a favorite uncle named Jakob who taught him mathematics as a boy. "Algebra is a merry science," Jakob once said. "We go hunting for a little animal whose name we don't know, so we call it *x*. When we bag our game, we pounce on it and give it its right name."[5]

Uncle Jakob's words stayed with Einstein for the rest of his life. They encapsulated his attitude toward mathematical and scientific problems, which to Einstein always seemed more like puzzles or games than like work. Consequently, Einstein could focus on his math studies all the serious concentration that most children reserve for play.

After first imagining himself running beside a light beam at age sixteen, Einstein spent the next ten years studying physics and pondering that same light-beam image. The reason Einstein had the patience to stick with the same "game" for ten years is that he had just as much fun "playing" it as he finally did solving it.

THE FLOW STATE

Psychologist Mihalyi Csikszentmihalyi wanted to know why so many people were so unhappy. Why, he asked, despite all the comforts, luxuries, and opportunities of our modern world, did people still "end up feeling that their lives have been wasted, that instead of being filled with happiness their years were spent in anxiety and boredom"?[6]

Csikszentmihalyi resolved to find out. For twenty-five years, he interviewed hundreds of people all over the world, from all walks of life—artists, athletes, chess masters, janitors. He asked them all to recall their happiest moments and to describe what brought those moments about. With remarkable uniformity, they all told him the same thing.

"The best moments," writes Csikszentmihalyi in *Flow: The Psychology of Optimal Experience,* "usually occur when a person's body or mind is stretched to its limits in a voluntary effort to accomplish something difficult and worthwhile. . . . Such experiences are not necessarily pleasant at the time they occur. The swimmer's muscles might have ached during his most memorable race, his lungs might have felt like exploding, and he might have been dizzy with fatigue—yet these could have been the best moments of his life."

Csikszentmihalyi called this sort of peak experience the flow state. It is the closest thing to heaven on earth. Csikszentmihalyi determined that flow occurs when we are absorbed in an activity that is neither too easy for us nor too difficult. If it's too easy, we become bored. If it's too hard, we become anxious. But if it is just right, the flow response will kick in. We then find ourselves in that same state of gentle concentration that Maria Montessori observed in children at play.

Flow Builds Brainpower

Scientists have found that pleasure is a key ingredient in building brainpower. We learn far more from mental exercises when we enjoy them.[7] Likewise, Csikszentmihalyi found that our minds grow in complexity the longer we remain in flow (see Figure 14.1).

Radio experts measure the clarity of a transmission by its signal-to-noise ratio. If the signal is much stronger than the noise, you can hear it clearly. If the signal is weak compared to the noise, you hear mainly background static (see Figure 14.2).

Figure 14.1 The longer you stay in flow—a state of pleasantly absorbed concentration—the smarter and more able you become. Suppose A in this diagram stands for Alex, a boy learning to play tennis. When he is just beginning (A1), volleying the ball back and forth over the net is an absorbing and intriguing challenge. Soon, however, Alex's skills improve, and he gets bored with endless volleying (A2). In order to achieve flow, he must increase the level of challenge, by playing against a more practiced opponent. At first, the opponent is so competitive, that Alex misses every shot. He becomes anxious and does not enjoy the game (A3). The only way for Alex to return to flow is to keep practicing until he is nearly as good as his opponent (A4).

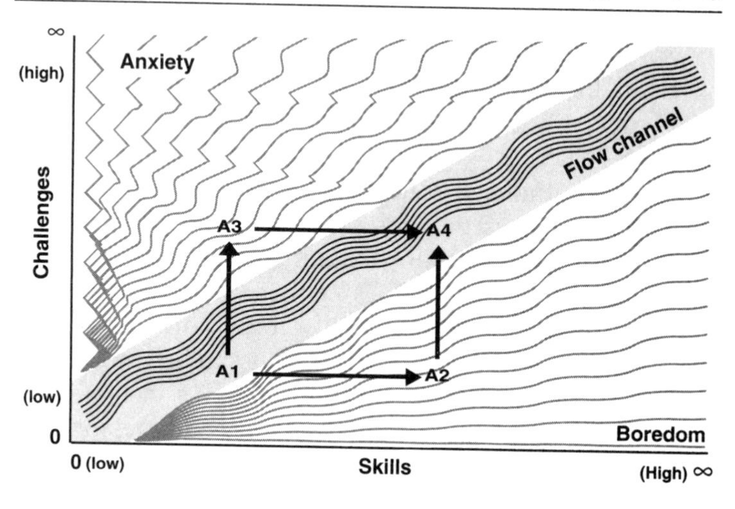

Flow increases mental intensity by raising the signal-to-noise ratio in our minds. When we perform a task that is too difficult, we find ourselves awash in feelings of shame, insecurity, and perhaps fear for our jobs or our grades. Such negative feelings create a constant and debilitating background noise that distracts from our work. Likewise, when a job is too easy, our minds wander just as erratically, through

Figure 14.2 (a) The signal is lost in background static with a poor signal-to-noise ratio. (b) A good signal-to-noise ratio allows the signal to stand out clearly.

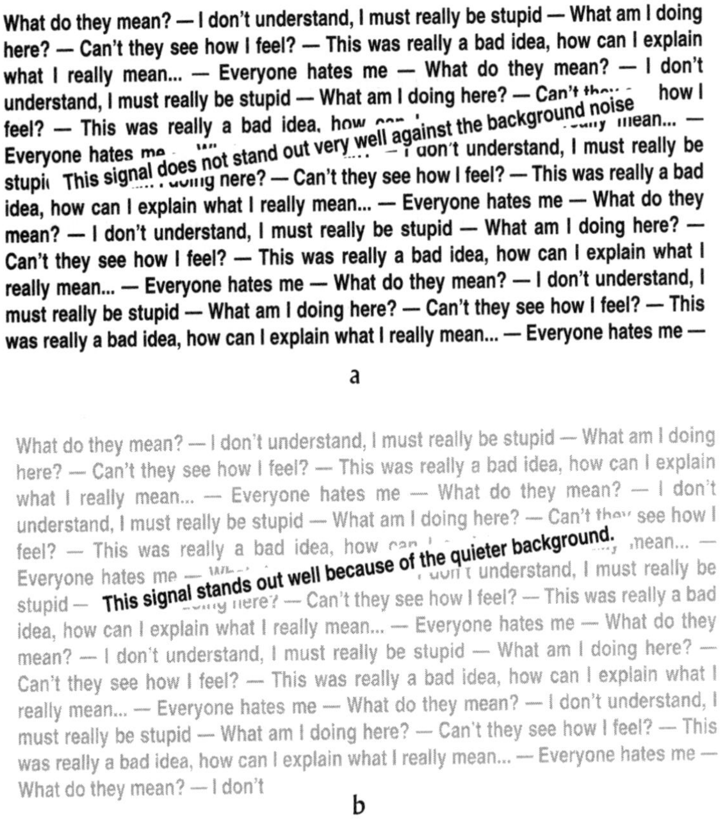

What do they mean? — I don't understand, I must really be stupid — What am I doing here? — Can't they see how I feel? — This was really a bad idea, how can I explain what I really mean... — Everyone hates me — What do they mean? — I don't understand, I must really be stupid — What am I doing here? — Can't they see how I feel? — This was really a bad idea, how can I explain what I really mean... — Everyone hates me — What do they mean? — I don't understand, I must really be stupid — What am I doing here? — Can't they see how I feel? — This was really a bad idea, how can I explain what I really mean... — Everyone hates me — What do they mean? — I don't understand, I must really be stupid — What am I doing here? — Can't they see how I feel? — This was really a bad idea, how can I explain what I really mean... — Everyone hates me — What do they mean? — I don't understand, I must really be stupid — What am I doing here? — Can't they see how I feel? — This was really a bad idea, how can I explain what I really mean... — Everyone hates me —

This signal does not stand out very well against the background noise

a

What do they mean? — I don't understand, I must really be stupid — What am I doing here? — Can't they see how I feel? — This was really a bad idea, how can I explain what I really mean... — Everyone hates me — What do they mean? — I don't understand, I must really be stupid — What am I doing here? — Can't they see how I feel? — This was really a bad idea, how can I explain what I really mean... — Everyone hates me — What do they mean? — I don't understand, I must really be stupid — What am I doing here? — Can't they see how I feel? — This was really a bad idea, how can I explain what I really mean... — Everyone hates me — What do they mean? — I don't understand, I must really be stupid — What am I doing here? — Can't they see how I feel? — This was really a bad idea, how can I explain what I really mean... — Everyone hates me — What do they mean? — I don't understand, I must really be stupid — What am I doing here? — Can't they see how I feel? — This was really a bad idea, how can I explain what I really mean... — Everyone hates me — What do they mean? — I don't

This signal stands out well because of the quieter background.

b

boredom. Once again, the clear signal of our work is swamped with noise.

People in flow, on the other hand, experience a profound state of absorption in which time freezes, anxieties fall away, and nothing seems to exist except the work at hand. All 126 bits per second of their conscious attention are filled. No room is left over for negative or distracting thoughts to creep in. At such times, thought becomes focused and orderly, like a laser beam.

A light bulb emits waves of light in random patterns and in many scattered frequencies. Such incoherent light is weak. It has no impact on the physical world around it. But if we bring order into that light beam, confining all the light waves to one frequency and aligning them in phase with one another, *voila!* We have a laser beam (see Figure 14.3). Such coherent light is far more powerful than its incoherent

Figure 14.3 Normal light (top) is incoherent. The light waves are random and out of phase. Laser light (bottom) is coherent. All waves oscillate in phase. Scientists have discovered that brain waves, too, can achieve laserlike coherence, especially when we experience sudden creative insights.

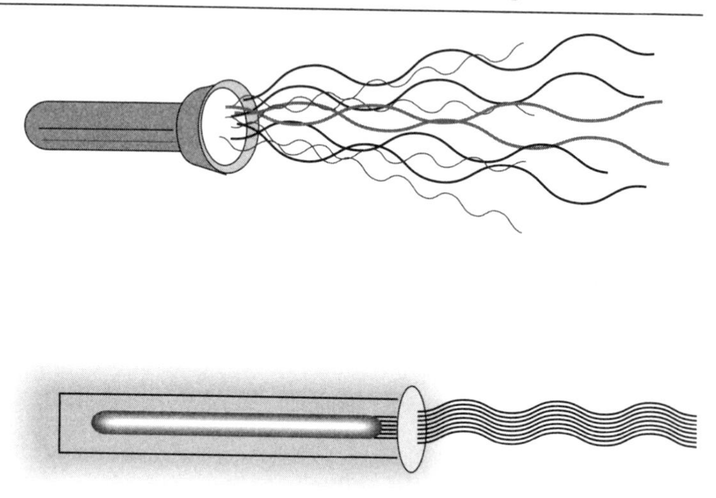

cousin. It can do everything from cauterizing a tiny injury on the retina to blazing a hole in an enemy satellite. A laser beam's power comes from its order and focus.

The mind, too, gains extraordinary power when it acquires order and focus. People in flow work with a level of speed, talent, and endurance that, at times, seems almost superhuman. In this condition, Mozart dashed off masterpieces in a single night, Babe Ruth hit his sixtieth home run of the 1927 baseball season, rock climbers in Yosemite National Park scale El Capitan, and surgeons perform marathon operations, realizing only when it's over what a magnificent feat they have achieved. This is the hyperproductive state that we at Project Renaissance call Creative Fire.

Pleasure through Feedback

Of course, flow's effects are seldom so dramatic. More often, we move in and out of flow without even realizing it. It can happen when you fall into a rhythm while practicing your tennis strokes against a wall. It can happen when you walk down the street and find yourself so absorbed by the beautiful, crisp sunlight on the buildings that you forget all your troubles. It comes when you are playing with your children or enjoying a quiet moment with your spouse. Virtually any stimulating activity that completely fills your 126 bits per second of conscious attention can put you in flow. Anytime you feel noise creeping into your mind in the form of worry, envy, anger, insecurity, and the like, you know that whatever you are doing at that moment is either too easy or too difficult for you. If you adjust your activity so that it fully absorbs all 126 bits, you will effectively block out the noise.

The chart in Figure 14.4 was devised by editors at *Success* magazine, based on the ideas of Dr. Csikszentmihalyi. It provides a reliable critical path for achieving and sustaining a high state of flow.

According to Csikszentmihalyi, feedback, more than any other factor, injects a gamelike fun into work and

Figure 14.4 The components of flow form a pathway to a high level of productivity.

I. Gamesmanship

Look at your task as a game.
Establish objectives, challenges
to be overcome, rules, and rewards.

II. Powerful Goal

As you play the game, remind yourself
frequently of the overriding spiritual, social,
or intellectual purpose that drives your work.

III. Focus

Release your mind from all distractions, from
within or without. Focus your entire
attention on the game.

IV. Surrender to the Process

Let go. Don't strive or strain to
achieve your objective. Just enjoy
the process of work.

V. Ecstasy

This is the natural result of the preceding four
steps. It will hit you suddenly, by surprise.
But there will be no mistaking it.

VI. Peak Productivity

Your ecstatic state opens vast reservoirs of
resourcefulness, creativity, and energy. Your productivity
and quality of work shoot through the roof.

learning. A good game always involves some way of keeping score. The score feeds back to the player, giving him a constant feel for how well his efforts are succeeding. We are far more motivated to shoot baskets if we keep track of the ratio of hits to misses. Completing a strenuous workout on the stair-step machine is far easier if you have a digital readout telling you each second how many calories you've burned and how many "floors" you've climbed.

Master Each Step

Mathematics is the least compromising of all subjects. Either you know the material cold or you don't know it at all. For that reason, math anxiety is the most common and crippling of all learning blocks.

Surprisingly, even this fearsome hurdle is easily vaulted by applying the principles of flow. Kumon Mathamatex, for example—a teaching system devised by Japanese educator Toru Kumon—has turned math from an ordeal into a delightful game for many students, U.S. and Japanese alike.

Kumon students work their way through a series of 4,400 worksheets that take them from basic arithmetic through calculus. Each student stays on the same worksheet until he can "beat the clock," getting a perfect score within a certain time period (usually 15 to 30 minutes). Instead of fretting over wrong answers, the student's entire focus is on breaking his previous speed record. Throughout the Kumon Mathamatex program, students remain in flow. They are constantly challenged to work faster and harder, but they proceed at their own pace, tackling only work that fits their abilities.

Conventional math students, in order to keep up with the rest of the class, must often push forward to the next lesson before they have fully mastered the last. This situation is like building a tower on a crumbling foundation. The higher the student rises, the more anxious he becomes, clinging for very life to his ever-more wobbly edifice. Feedback, in the form of grades, becomes more and more discouraging. Anxiety creeps in. Flow becomes impossible.

Kumon Mathamatex students, on the other hand, never move on to a new lesson until they can perform the last lesson perfectly, at high speed. They take on each new lesson bursting with the magisterial confidence that comes only with consummate mastery.[8]

Love the Process

During the production of *Fantasia,* Walt Disney exhorted his animators with the slogan "This is not the 'cartoon' medium. We have worlds to conquer here."[9] For Disney, animation was more than a branch of show business. It was a miraculous medium that would someday expand human expression to the point where we could bring our wildest imaginings to life virtually at will. "What I see way off is too nebulous to describe," he once said of the industry's future. "But it looks big and glittering."

Disney's vision is already taking shape in the new fields of virtual reality and digital special effects. Unfortunately, these advances came thirty years too late for Walt Disney. Like Moses, Disney spent his life journeying toward a Promised Land he was destined never to see. Should we pity him? Not according to Disney. While he was alive, he accepted with perfect grace the rules of the "mortality game."

"We, the last of the pioneers and the first of the moderns, will not live to see this future realized," he said. "We are happy in the job of building its foundations."[10] Like all great geniuses, Disney found his joy in the *process.* His lofty goals functioned mainly as an excuse to keep plugging away at the work he loved.

THE SLIGHT EDGE

As president of The People's Network (TPN), a fast-growing satellite TV network based in Irving, Texas, and devoted to self-improvement programming, Jeff Olson describes a success philosophy he calls the Slight Edge. "Winners have an edge over everyone else," he writes in *Success* magazine. "But it's only a *slight edge*. The difference between success and failure often lies in daily actions that are easy to do—and also easy not to do. . . . The slight edge is always at work, either for or against you."[11]

Over time, says Olson, our minute daily habits build up to gigantic achievements—or gigantic failures. He likens this process to a single water hyacinth that doubles each day. After 15 days, only a tiny cluster of water hyacinths drifts in the pond. On day 29, the pond is half covered.

"But on day 30," says Olson, "the entire surface is one giant mass of water hyacinth."

Einstein's Slight Edge

After failing to obtain an academic appointment and settling for a lowly job as a patent examiner, Einstein could easily have succumbed to discouragement and domestic pressures. Visitors to Einstein's apartment in those years expressed horror at the squalor of his life.

"The door of the apartment was open," wrote David Reichinstein, a former college professor of Einstein's, "to allow . . . the washing hung up in the hall to dry. I entered Einstein's room. He was calmly philosophic, with one hand rocking the bassinet in which there was a child. In his mouth, Einstein had a bad, a very bad, cigar, and in the other hand an open book. The stove was smoking horribly. How in the world could he bear it?"[12]

In such surroundings, it would have been easy for Einstein to put aside his physics studies. He could have salved his conscience each day with the excuse, "Things are a bit hectic right now. I'll get back to my studies tomorrow." After a few years of such procrastination, Einstein would no longer have felt the need for excuses. His light-beam problem would have been forgotten.

But that is not what happened. Einstein had a slight edge. His edge was that he *enjoyed* studying physics. He enjoyed it so much that, no matter how busy he was, he always found time each day to pursue his passion. Over ten years' time, the single water hyacinth of Einstein's thought experiment with the light beam grew to fill an entire pond. It came as a complete surprise to Einstein when his years of patient plodding suddenly exploded onto front pages across the world. In his own mind, he had simply been pursuing a much-loved hobby, a little bit every day.

THE AUTOTELIC IMPERATIVE

Dr. Csikszentmihalyi would say that Einstein's physics studies were an *autotelic* activity. *Autotelic* comes from the Greek root *auto,* meaning "self," and *telos,* meaning "goal." An autotelic activity is a goal in itself. We pursue such activities not in the hope of some future gain but for the sheer pleasure of doing them.[13]

The desire for good grades or a better job is poor fuel for study. Only when you pursue knowledge autotelically will you have the drive and the patience to master a skill as thoroughly as Einstein mastered physics, Mozart music, or Disney animation.

It's Up to You

How you go about finding pleasure in learning is a highly individual affair. Each of us must find his own knack for injecting fun into learning. No single method will work for everyone. Many people, for example, find rote memorization to be a painful ordeal, as it long was for me. For coauthor Richard Poe, it has proved to be extraordinarily autotelic.

Having spent five years studying Latin in high school and college, Richard had long been frustrated by his poor comprehension. He resolved recently to build his fluency by memorizing lengthy texts in Latin. Within a few months, Richard had memorized the entire first chapter of the Latin Vulgate Bible.

Unlike previous self-study projects, which he took up for a few weeks and then abandoned for lack of time, Richard found that he had no trouble sticking to his Latin regimen. For one thing, it cost him no extra time. He could glance at a line in the morning and spend the rest of the day repeating it in his head at odd moments. Once-tedious activities, such as standing in line at the bank or riding a crowded subway, became golden opportunities for clearing his mind and practicing his Latin. Richard used to feel guilty about the time he wasted lying in bed waiting to fall asleep. Now he could recite Latin verses during that idle time.

It is hard to say whether Richard will ever master the Latin language through this method, but that goal has become irrelevent. For Richard, the sheer delight of reciting hundreds of lines of Latin verse from memory—and watching his inner library grow with each passing week—is sufficient reward. Quite by accident, Richard has stumbled upon a method for permanently increasing the order and discipline of his mind. He spends far more time these days absorbed in challenging mental exercise, and far less time being bored or anxious. According to Dr. Csikszentmihalyi's theory, such a prolonged state of flow should, over the years, steadily increase the complexity of Poe's thought processes.

METASTRATEGY

Many readers no doubt would consider the memorizing of Latin scripture a refined form of torture. Each must seek his own path to *autotelic discipline*. I cannot tell you what your method should be, but I can provide some hints on how to find it.

Most of the techniques you have learned in this book are metastrategies, that is, strategies that can be used to develop other strategies. The boy I described in Chapter 1 used my Borrowed Genius technique to discover an amazingly effective method for hitting a baseball. Deep in your unconscious mind lie unique strategies that will allow you to hone and develop your left-brain skills with a daily passion. You can find them through a process I call Toolbuilding.

The Toolbuilding Process

Virtually any Project Renaissance method can be used as a Toolbuilder, a metastrategy for devising other learning strategies. Below are the five Toolbuilding applications I have found to be most effective. Toolbuilding continues to startle us year after year with a cornucopia of new methods for training and learning. Its astonishing fertility has been one of the biggest surprises at Project Renaissance. We trust it will serve you as bountifully.

Method 1: Straight Image Streaming Ask your greater consciousness what methods will best enhance your ability to achieve flow and to grasp such-and-such subject matter. Then Image Stream in the normal fashion, describing your images aloud to a partner or tape recorder, and interpret the results.

Method 2: Thresholding In your Image Stream, imagine an Answer Space, concealed perhaps in a secret garden behind a high wall, as described in Chapter 5. In that Answer Space lies a clue to the new learning method you seek.

Method 3: Borrowed Genius Put on the head of a genius who is uniquely qualified to invent new learning methods. Then ponder the question in your guise as that genius. You can also use the variant of Alternate Self or Parallel World, as described in Chapter 8. Simply imagine that your own counterpart on a parallel planet is the greatest living expert on devising new methods of learning.

Method 4: Instant Answer Close your eyes and look into your Image Stream to see what pictures are there right now. Do this three times and compare the images. The common elements of all three picture answers will reveal something about the new technique you are searching for.

Method 5: Advanced Civilization For some reason, this technique has been the most fruitful of all Toolbuilders used in our seminars. Simply use some imaginative device, such as the Space/Time Transporter in Chapter 8, to whisk you away to an advanced, futuristic civilization. Your civilization should be peopled by biologically normal humans, rather than exotic aliens. The reason they need to be human is so that you can learn from them secrets of ingenious thought that will also apply to an earthling like yourself.

However, these seemingly normal humans are far from ordinary. In the futuristic world you imagine, every ten-year-old is a better violinist than Heifetz, a more profound thinker than Einstein, or a better playwright than Shakespeare—depending on the area of expertise you seek to improve.

Let the Image Stream take you through an experience in which you somehow learn the secret of that advanced civilization, the reason why its people are so much more advanced in that particular subject or skill than the greatest scientists, artists, or virtuosos of our present-day earth.

Put on the head of someone in that civilization, such as a child, and find out what is done to that child, how he is treated and raised, and what experiences he has that stimulate his talents and intellect so powerfully.

The Toolbuilding Tradition

I hope that readers will take their Toolbuilding seriously. In some ways, it is the most important technique in this book. *The Einstein Factor* itself and all the methods in it are products of years of Toolbuilding, as is the entire field of creativity enhancement.

Probably no one is more respected in that field than Dr. Sidney J. Parnes, a close associate of the late Alex

Osborn, the inventor of brainstorming. It would have been very easy for Parnes to rest on his laurels, after helping to devise the famous Osborn/Parnes methods of creative problem-solving 50 years ago. Yet, year after year, he has continued to reinvest those original metaskills in a never-ending quest for new and better methods. Parnes' tireless commitment to the Toolbuilding ideal has made him the world's leading expert on creativity and his Creative Education Foundation the preeminent institution in that field.

Parnes' seminal work is *Visionizing: State-of-the-Art Processes for Encouraging Innovative Excellence* (Buffalo, NY: Creative Education Foundation, 1988). For an overview of Parnes' contribution, I recommend the *Source Book for Creative Problem Solving: A 50-Year Digest of Proven Innovative Processes* which Parnes edited (Buffalo, NY: Creative Education Foundation, 1992).

LIFE IS YOUR MASTERPIECE

"Is there not a certain satisfaction," Einstein once wrote, "in the fact that natural limits are set to the life of the individual, so that at its conclusion it may appear as a work of art?"[14]

These words are peculiarly poignant coming from Einstein, for when he died at age seventy-five, he left his greatest work undone. By ordinary thinking, the forty years Einstein spent searching for the Unified Field Theory were wasted. When that problem is finally solved, the glory will go to another.

Yet Einstein went to his death content, for he had richly enjoyed each moment spent in pursuit of that elusive Grail. When our last day comes, will we too note with satisfaction the many years we spent in flow, reveling each day in joyous tests of our ability? Or will we have to face the dismal truth that for most of our years we oscillated nervously between anxiety and boredom?

Whether you are eighteen or eighty, the time is now to take on that grand project of your dreams. Don't worry whether you'll have the time to complete it. Master the skills day by day. Enjoy them to the fullest. Silently, ploddingly, and unobtrusively, the slight edge will go to work in your life. Through it all, may The Force be with you.

CHAPTER 15

THE GENIUS MEME

Who can imagine Egypt without her pyramids? If not for these famous monuments, most ordinary people would have little idea who the Egyptians were or what they accomplished. The world of the pharaohs would seem as obscure to us as India's lost city of Mohenjo-Daro.

Yet the invention of the pyramid was in no way inevitable or intrinsic to the Egyptian soul. Indeed, no such structures would have existed in Egypt had it not been for the genius of a single man.

THE FIRST GENIUS IN HISTORY

Five thousand years ago, the pharaohs of Egypt were buried in squat, mud-brick structures called "mastabas." Then a court architect named Imhotep had a better idea. Instructed to build a royal tomb for the pharaoh Djoser, Imhotep piled an incredible 850,000 tons of limestone into a structure soaring 200 feet above the desert.[1] Nothing like it had ever been built. It was not only the first pyramid in history, but the first high-rise stone edifice of any sort.

Imhotep was nothing less than history's premier genius. After 4,700 years, his legacy lives on. From the glass pyramid in the courtyard of the Louvre to the Transamerica Pyramid that dominates the San Francisco skyline, modern architects continue to imitate Imhotep's handiwork. Indeed, each time modern builders lay one stone upon another, they honor the man who archaeologists agree virtually invented the craft of stonemasonry.

MEMES, NOT GENES

Einstein once wrote that "all the valuable things, material, spiritual, and moral, which we receive from society can be traced back through countless generations to certain creative individuals."[2]

He was talking about people like Imhotep. What makes such creative individuals special? Why does one caveman carve a dagger from flint, while his fellow hunters use sticks? Why does one cavewoman plant seeds in the ground, while her companions scrounge for roots and berries? Geneticists have long sought a "genius gene" that might explain such exceptional insight. Yet the growing weight of evidence drives us to look elsewhere. The secret of genius appears to lie not in our genes, but in our *memes*—those patterns of thought, habit, and emotion woven into our minds by the people and events around us.

SURVIVAL OF THE FITTEST

If a lioness finds a cheetah hunting in her territory, she will usually try to kill it. Moreover, if the lioness happens upon a den of cheetah cubs, she will systematically murder each cub by clamping her jaws over its face until it suffocates, leaving the bodies for the mother cheetah to find.

As Howard Bloom points out in *The Lucifer Principle,* such ruthless intolerance typifies the natural world. One set of genes is always trying to overpower and supplant some other set of genes. Indeed, individual creatures often seem to be little more than expendable pawns in the power struggle between competing gene pools.[3]

The Selfish Gene

According to Bloom, this age-old drama dates back 4 billion years. At that time, the closest things to life on earth were microscopic entities called genes floating around in the primordial ooze. To gain a competitive edge, some of these genes grouped together into colonies, called DNA. Then other genes did the same, in order to keep up. The arms race was on.

In their quest for more efficient ways of killing, devouring, parasitizing, and otherwise overpowering their rivals, these gene colonies evolved, over billions of years, into ever more complex forms of animal and plant life. According to this theory—which Bloom cites from the 1976 book *The Selfish Gene* by Oxford zoologist Richard Dawkins—we humans are little more than the latest weapon system developed by our genes in their relentless campaign to dominate the natural world.

The Secret Weapon

One day, a particularly enterprising set of genes developed a secret weapon—intelligence. This added a whole new dimension to the genetic arms race. In the past, new weapons such as teeth, claws, antlers, and poisonous venom had to develop slowly over millions of years. The secret ingredient of intelligence allowed entire gene pools to shift their survival tactics overnight.

Some chimpanzees, for example, have learned to use twigs for fishing out termites from a hole, a skill that gives a chimp colony a competitive advantage over its neighbors, who may be starving for lack of termites. No chimpanzee is born with this termite-fishing skill. It does not pass on to the next generation through the genes. Instead, it can be spread directly from one chimp brain to another through a teaching process, pervading a chimp population almost overnight. For this reason, Dawkins would say that termite fishing is a *meme*.[4]

AN EPIDEMIC OF IDEAS

Dawkins defined a meme as a set of ideas that can replicate itself like a virus and spread from one brain to another. The viral analogy is apt, for memes have been found to diffuse through a population in a manner that is mathematically similar to the epidemiology of plagues.[5]

In 1919, a new strain of influenza virus succeeded in killing more than 21 million people in less than a year. Memetic action is sometimes slower, but it is equally potent. As Bloom observes, it took only 300 years for the meme of Christianity to overrun the Roman Empire and less than 130 years for the meme of Communism to leap from the pages of *Das Kapital* to "infect" 1.8 billion people across the globe.[6]

The Amoral Meme

Like the selfish gene, the meme has a blind, unthinking drive to reproduce at any cost. Memes recognize neither good nor evil. The memes of Nazism, Islam, hula-hoop mania, double-entry bookkeeping, and the Jeffersonian principles of life, liberty, and the pursuit of happiness all spread with equal virulence when they first infected human brains. Memes are just as efficient at spreading bad or trivial ideas as good ones. Whether they help or harm mankind depends entirely on how we apply them.

THE JURASSIC PARK PRINCIPLE

Geneticists today possess most of the skills necessary to resurrect long-extinct dinosaurs. They lack only the right genetic code. If we could somehow find a complete strand of dinosaur DNA, we could theoretically clone a dinosaur that would be biologically identical to its ancestors who lived 60 million years ago. As in Steven Spielberg's movie *Jurassic Park*, the cloned dinosaur would have the same diet, the same mating rituals, the same migratory habits, and probably the same temperament as its long-dead forebears. Given a suitable mate, it would lose no time in setting up housekeeping and spawning a new generation of dinosaur babies.

We know this would happen because the very existence of dinosaurs in past millennia proves that their genetic code works. Provided that we have accurately duplicated that code in every detail, we can expect that the dinosaurs it produces will thrive and procreate as vigorously today as in the Jurassic period (assuming, of course, that we have placed them in a Jurassic-like environment).

Ideas from the Past

Extinct ideas can also be resurrected by faithfully reproducing their memetic codes. We know, for example, that the language of the ancient Hebrews works, because it survived for many centuries as a living tongue. When the Hebrew meme was revived as the official language of modern Israel in the 1940s, it caught on quickly, even though the language had been dead for over 2,500 years.

Attempts to introduce the Esperanto language, on the other hand, failed miserably. Esperanto was created by modern-day utopians who hoped to unite the world under a common tongue. Unlike the genes of *Tyrannosaurus Rex*, the meme of Esperanto had not been forged in the tooth-and-claw struggle for survival. It was a pure invention. In the real world, it simply didn't work.

Genius Breeds Genius

We know for a fact that there is a Genius Meme and that it works, because this meme has infected human civilization many times in past ages. Historians have long noted that genius never seems to appear alone. It generally sweeps upon humanity in a great wave of hyperachievement, later recorded in history books as a Golden Age or Renaissance.

When Imhotep died, the Egyptians did not revert to their past dullness. Infected with the meme of pyramid building, succeeding pharaohs poured all their wealth and power into raising bigger and better pyramids, culminating in the Great Pyramid of Khufu, a gargantuan edifice 150 meters high composed of 6.5 million tons of limestone.[7] The five largest and most famous pyramids in Egypt were all built within 100 years of Imhotep's death.

In the fifth century B.C., the single Greek city-state of Athens produced the philosophy of Plato and Socrates, the sculpture of Phidias, the statecraft of Pericles, the plays of Sophocles and Euripides, and the science of Aristotle—all within a single lifetime. Two thousand years later, another city-state called Florence experienced a Renaissance that numbered among its prodigies such giants as Michelangelo, Leonardo da Vinci, Botticelli, Mirandola, and Fra Filippo Lippi.

PROJECT RENAISSANCE

The *Jurassic Park* Principle assures us that we, too, can experience a Golden Age. The only requirement is that we start with the right memetic code. Unlike Esperanto, the Genius Meme has a long and vibrant track record. It has worked many times throughout history and will surely work again.

Through Project Renaissance—a nationwide network of accelerative-learning enthusiasts—I have, for some years, been inoculating society with my version of the Genius

Meme, using every known method of transmission from books, tapes, seminars, and software to good old-fashioned word-of-mouth. Someday soon, I hope that we will reach critical mass and begin to see profound changes in society.

The Structure of the Genius Meme is depicted in Figure 15.1. Four of the components are already familiar to you from past chapters. Thin Boundaries, as we know, can be attained through the daily practice of Image Streaming and related techniques, as can the habit of Original Observation. The most appropriate Autotelic Discipline for you—along with the Strong Left-Brain skills it engenders—is best sought through Toolbuilding.

That leaves Primitive Drive and Noble Spirit, two qualities that seem, on the surface, to be polar opposites. In fact, they are two sides of the same ingenious coin.

PRIMITIVE DRIVE

Few figures in history are more noble than Abraham Lincoln. Yet his drive to succeed was so strong that contemporaries often found him pushy and self-serving. His old law partner, William Herndon, remarked somewhat acidly that Lincoln's ambition was "a little engine that knew no rest" and that he was "always calculating and planning ahead."[8]

Sigmund Freud would have attributed Lincoln's drive to a powerful but sublimated libido. Freud believed that human beings have a unique gift for rechanneling their sexual desire into an equally compelling urge to rear skyscrapers, write books, win Civil Wars, or sail unexplored seas.

The great motivator Napoleon Hill agreed. "Sex energy is the creative energy of all geniuses," he wrote in 1937. "There never has been, and never will be a great leader, builder or artist lacking in this driving force of sex."[9]

Figure 15.1 Working from fossilized DNA, scientists in the movie *Jurassic Park* created living, breathing dinosaurs. Methods of thinking can also be resurrected from the past. They are preserved, not in DNA, but in memes—self-replicating patterns of thought and behavior that can spread from one mind to another like a viral infection. In order to bring back the golden days of the Italian Renaissance or fifth-century Athens, we must infect modern society with the same Genius Meme that thrived in the brains of Socrates and Leonardo da Vinci. It probably looked something like this.

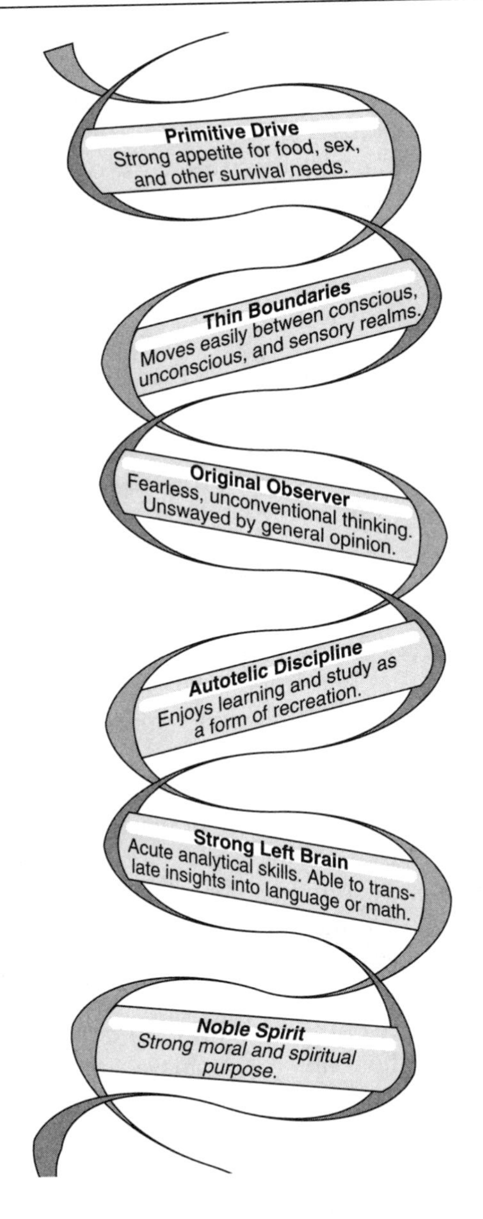

Hill listed Lincoln alongside William Shakespeare, George Washington, and a host of other luminaries whose "highly sexed" nature he believed had fueled their success. This did not mean that they were womanizers. Quite the contrary. Physical sex, said Hill, tended to dissipate one's creative drive. The habit of many men to "sow their wild oats" in youth was, in Hill's opinion, largely responsible for their failure to distinguish themselves until after the age of forty. Great geniuses, Hill averred, succeeded precisely to the extent that they redirected their sexual desire *away* from physical expression and "transmuted" it into more fruitful ambitions.

Gorging and Fasting

The sex drive, of course, is not the only primitive stimulant to which the mind is subject. Animals in the wild are driven just as powerfully by their hunger for food.

I have long noted an unusual frequency of eating disorders among my fellow MENSA members, ranging from obesity to anorexia and bulimia. Perhaps this observation indicates some connection between appetite and intelligence. For centuries, saints and mystics have sought spiritual transcendence through fasting. Can it be that, like Freud's sexual "sublimation" and Hill's "sex transmutation," fasting serves to rechannel the primitive drive of hunger into higher pursuits?

In fact, modern brain science suggests that it is. Experiments have shown that many forms of arousal—including hunger—help prime our brains for ingenious thought. When hunting grew lean, our cave-dwelling ancestors had to find innovative new food sources. Those who failed to become smarter tended to starve. They seldom lived long enough to procreate, become our forebears, and pass their traits onto us! With each generation, evolution thus reinforced our ability to achieve temporary IQ spikes.

Primed by Arousal

In Chapter 5, we described the experiment in which Walter Freeman and Christine Skarda measured the chaotic brainstorms that cascaded through rabbits' brains when they sniffed a familiar scent.

In order for such a burst of perception to explode across the brain, the researchers found, the brain cells must first be electrically primed by physically *arousing* the rabbit through sexual desire, hunger, thirst, or even physical threat. Such arousal increases the gain—a measure of electrical potential—between the rabbit's brain cells, making the brain hyperreceptive. Apparently also in humans, such arousal and receptivity encourage the bursts of perception by which we form ingenious *gestalten*.[10]

"ACTS OF INTELLIGENCE"

When we feel threatened, our hearts race and our breathing speeds up. Later, when the danger is past, our bodies return to normal. Dr. Luiz Machado, an accelerative learning pioneer and Professor of Modern Languages at the State University of Rio de Janeiro, believes that ingenious thought follows a similar sequence.

"People are not intelligent the whole time," says Machado. "Acts of intelligence are produced when the neuroglandular circuits are activated in order to meet specific needs."[11]

As we noted in Chapter 13, our brains respond to a 10-minute treat of Mozart's *Sonata for Two Pianos in D Major,* K. 448, by immediately becoming 8 to 9 IQ points "smarter" and then returning to normal about 15 minutes later. This temporary "act of intelligence" may mirror the emergency IQ spikes that Machado believes were programmed into our primitive brains in the wild.

The Limbic System

According to Machado, "acts of intelligence" are instigated by the limbic system, a set of structures located beneath the cerebral cortex in the deeper or more primitive midbrain area (see Figure 15.2). Scientists believe that the limbic system governs emotion, as well as such basic drives as aggression, hunger, sex, and parental instinct.

The limbic system also acts as a high-speed processor of sensory impressions. It sorts and analyzes input from all five senses, fires off the results to appropriate areas of the cortex, and then reassembles the data that comes back for further analysis. Through this high-speed feedback loop, repeated over and over in a matter of split seconds, the limbic brain forms ever more complex *gestalten* of the world around us.[12,13]

Figure 15.2 Genius is linked to strong appetites for food, sex, and other survival needs. Such primitive drives originate in the limbic system (shaded).

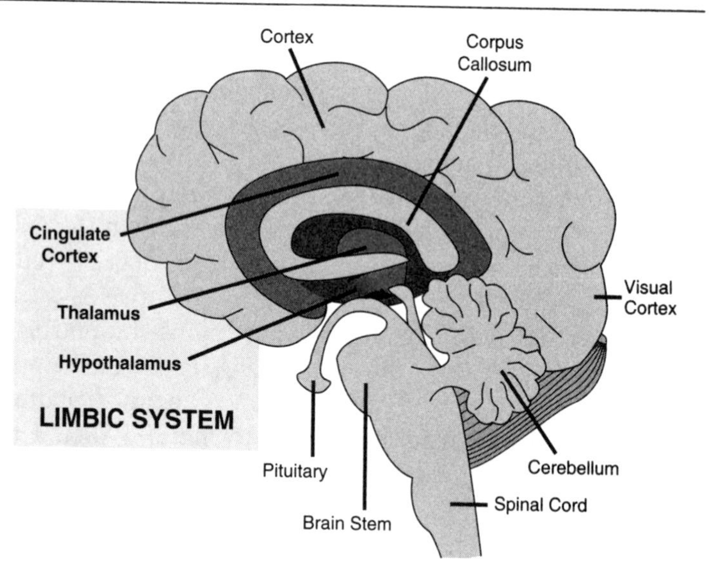

At any time during the sequence, the limbic system may seize upon a particular *gestalt* and initiate emergency action. It might, for example, suddenly realize that the lovely pattern of black and orange stripes in the trees is really a hungry tiger about to pounce. At such times, the limbic brain acts without bothering to consult the cerebral cortex. It spurs us to take defensive action even before we consciously recognize the danger.

We noted in Chapter 1 that the conscious mind can process only 126 bits of information per second and only 40 bits per second of human speech, yet our senses receive up to 10 million bits of input per second.[14] All this unconscious input is sorted, processed, and—wherever necessary—acted upon almost instantaneously in the limbic brain, which works approximately 80,000 times faster than the conscious cerebral cortex.[15] The limbic system's superfast processing enables a kangaroo rat to jump to safety fractions of a second before a diving eagle strikes.

Machado suggests that the limbic system's primary function is what he calls S.A.P.E.—a Portuguese acronym for Self-Preservation and Preservation of the Species (*Sistema de Auto-Preservação e Preservação da Espécie*). Drought, famine, predators, and declining breeding populations all threaten survival, either of individuals or of the species. Faced with such challenges, animals in nature are driven to their utmost exertions of strength, endurance, and creativity.

In its role as Guardian Angel, the limbic system will resort to any expedient that will ensure survival. According to Machado, our rational intellect, housed in the cerebral cortex, is but one of many emergency backup systems that the limbic brain can deploy or bypass as needed.

Thus, if a chimp is confronted with a log filled with juicy but inaccessible termites, the limbic brain might spark an "act of intelligence" to help the chimp figure out how to reach them. But if a leopard springs upon the chimp catching it unawares, the limbic brain will activate more primitive "fight or flight" reflexes that require no intelligence at all.

NOBLE SPIRIT

Humans tend to look upon primitive drives as innately more callous and cruel than our "higher" faculties of reason, compassion, and self-sacrifice. In fact, our higher faculties are governed by S.A.P.E., as well.

Einstein himself once pointed out that morality contributes to species survival and used the virtue of truthfulness as an example. "Lying destroys confidence in the statements of other people," said Einstein. "Without such confidence, social cooperation is made impossible or at least difficult. Such cooperation, however, is essential to make human life possible and tolerable."[16]

To preserve the species, many creatures will readily sacrifice their own lives. Salmon swarm upriver to spawn, braving punishing rapids and hungry bears, only to die of exhaustion shortly after mating. Mother birds will attack stalking predators many times their size to defend their chicks, and worker bees will sting to protect the hive even though the stinger will rip out their insides when it detaches from their bodies.

Generosity, compassion, devotion to duty, self-sacrifice, and romantic love all have their analogues in the animal world. It is no coincidence that these refined emotions are generated in the same limbic system that spurs us to eat, copulate, and, if necessary, kill. Our instinct for group survival is a powerful force that drives us to look past our narrow self-interest.

THE LONG VIEW

"In all ages," said William James, "the man whose determinations are swayed by reference to the most distant ends has been held to possess the highest intelligence."[17]

Distant indeed were the ends that inspired some primordial fish to venture upon the land and become an amphibian. Distant were the goals of the first protobird that spread its half-formed wings and leapt from a precipice. From the dawn of time, geniuses—human or otherwise—have led the vanguard of evolution, driven not by individual goals, but by the wordless, limbic impulse to advance the species.

WE ARE EVOLVING

Charles Darwin believed that a species advances only through "natural selection"—the survival of the fittest. Yet most geneticists today recognize that other forces are at work. In modern society, advanced medicine and social services shield most people from the grim consequences of natural selection. Yet evolution moves implacably forward.

Today, for example, more and more people are being born without third molars, especially among Eskimos and Europeans. More people are also appearing with three roots on their first molar, rather than the usual two. Such quirky changes seem to have little impact on who lives or dies, yet they proceed relentlessly ahead as if driven by some inner genetic clock.[18]

Unlimited Potential

At the inception of modern track and field events in the nineteenth century, it was considered physically impossible for a human being to run a mile in 4 minutes. Decade after decade, the greatest runners on earth consistently fell short of this milestone. Then, on May 6, 1954, Roger Bannister stunned the world by running the mile in 3 minutes 59.4 seconds. Since then, 4-minute miles have become routine.

Similar phenomena have occurred in the intellectual realm. For centuries, the greatest chess masters in the world tested their mettle by playing blindfolded. It was long believed that three blindfolded games at once marked the limit of human capacity. Then, in 1933, Alexander Alekhine successfully played thirty-two blindfolded games simultaneously. Later grand masters left Alekhine's record in the dust. When Koltanovski set the current record, playing fifty-six blindfolded games at once in 1960, he won fifty and finished the rest in a draw.[19]

Memetic Evolution

These advances were sparked by memes. Only the meme of athletic competition would induce a runner to push and train himself for years to the point of breaking the 4-minute mile. Only the meme of chess would give people any reason or desire to exercise their memories to the point of recalling fifty-six games at once. Once such breakthroughs are made, the meme is amended, and each succeeding generation performs the same task with less difficulty.

In the 1930s, Nazi geneticists imagined that the only way to improve the human species was through selective breeding, an assumption that led inexorably and tragically to the wholesale murder of anyone possessing "bad" or undesirable genes.

In fact, breeding is over 10 million years out of date as an efficient engine of human evolution. Since the development of the cerebral cortex and its unique capacity to form and transmit ideas, memes, not genes, have provided the most powerful stimulus for change.

SMART DRUGS

Of course, many promising methods for enhancing intelligence are physiological, not memetic. These techniques improve the hardware into which the memetic software is downloaded. I refer especially to the exciting new research in "smart drugs"—vitamins, herbs, amino acids, and various chemical compounds which some researchers claim can increase memory and intelligence.

Stimulated by the right memes, human beings will emit intelligence-building neurochemicals even without the intervention of smart drugs. Through this self-reinforcing feedback loop, the brain guides its own physical evolution over time. Still there is nothing wrong with giving evolution a jump start now and then, and that is what smart drugs are all about.

Due to the limited space in this book, I have chosen to focus only on those techniques with which I have long experience and expertise—the memetic ones. Readers would be well-advised, however, to investigate smart drugs on their own. This research is at the forefront of the accelerative learning field and early results have been quite impressive.

I recommend *Smart Drugs and Nutrients* by Ward Dean, M.D. and John Morgenthaler (B&J Publications, 1990, PO Box 483-905, Santa Cruz, CA 95061) and *Life Extension: A Practical, Scientific Approach* by Durk Pearson and Sandy Shaw (Warner Books, New York, 1984).

CHOOSE YOUR MEMES

In an anonymous article signed with the pen name "Keith Henson," a writer in the *Whole Earth Review* coined the term "meme-oids" to describe people so deeply in the grip of a meme that they lose all common sense. Religious and political fanatics of all stripes generally fall under this category. The critical weakness of a meme-oid is that he loses the capacity to choose his own memes. Instead, the memes choose him, and direct him to do their will.[20]

Don't be a meme-oid. By judiciously choosing your memes—and discarding those that get out of hand—you can engineer your own consciousness. History offers many examples in which the astute selection of memes greatly advanced human evolution.

In Renaissance Europe, for example, authorities in certain Italian cities did their descendants a tremendous service by authorizing a switch from Roman numerals to the new Arabic system. Before then, you had to be a genius to work out the simplest computations. Try multiplying 24 times 37. Most people today learned to do this in third grade. But if you try to multiply XXIV times XXXVII, you're in for a real challenge. The astute adoption of the meme of Arabic numerals vastly accelerated the capacity of human thought in a large portion of the world.

ACTIVATE YOUR GENIUS MEME

Having read through this book, you now possess the basic code for the Genius Meme and the tools to activate it. You can use that meme to resurrect in your own mind the Renaissance mindset that drove Socrates, Imhotep, and Leonardo da Vinci.

At the root of the activation process is your daily practice of Image Streaming. Once you have opened your mind to the subtle messages of the right brain and limbic system, the other components of the Genius Meme will fall into place naturally. Your own perceptions will guide you on your way, showing you, at every step, what to do next.

DESCRIBE ALOUD!

I cannot count the number of times over the years that readers of my past books have approached me and said, "I tried Image Streaming and it didn't work for me."

In such cases, I always ask, "Did you describe your images aloud to a live partner or to a tape recorder?"

This invariably brings a puzzled expression to the complainer's face. "Neither," he will finally admit. "I thought I could just do it in my head."

The attentive reader will remember that, in Chapter 3, I referred to such silent Image Streaming as the Insomniac's Special. It is good mainly for lulling yourself into a profound theta state, commonly known as *sleep*.

No portion of the technique is so often neglected as describing aloud to a partner or a tape recorder. Yet, none is more crucial to its success. Some readers may argue that they have already obtained good results from silent Image Streaming. Yet, they can vastly increase those benefits simply by following the full procedure.

TAKE THE LEAD

Just as the human brain uses chaos and disruption to break through to higher and more complex gestalten, so society needs a little disruption now and then to jar it loose from complacency and steer it toward a higher plane. Left to their own devices, our institutions tend to stagnate. For the sake of efficiency, our schools and workplaces trim us into sad little privet hedges, lined up in a neat row. Perhaps that is why it takes courage and more than a little self-sacrifice to walk the path of genius. Teachers, bosses, and community leaders will not always reward you for dodging the cruel hedge-clippers and spreading your branches to the sky.

We were not meant to be privet hedges. We were designed to grow tall like the mighty sequoia. Attend closely to the subtle messages from within. Follow the nobler promptings of your limbic brain. In a shorter time than you think possible, you will take your place in the vanguard of evolution, where you will find the deepest and most lasting satisfaction this life has to offer.

AUTHOR'S NOTE

Accelerative learning is still in its infancy. Poorly funded and largely neglected by established research institutions, it remains to this day a field where people in all walks of life, armed with little more than open and able minds, can still make a major impact.

Project Renaissance is designed to help specialists and non-specialists alike play a vital role in the ongoing Brain/Mind revolution. It is a network of men and women who share the goal of enhancing human intelligence, especially through the practice of Image Streaming and other techniques set forth in *The Einstein Factor.*

Through Project Renaissance, you can organize local Image Streaming support groups, network with leaders in brain/mind research, participate in workshops and seminars around the country, and keep abreast of breaking developments in the field through our newsletter.

For more information, send a self-addressed, stamped envelope with at least 64¢ postage to:

Project Renaissance
PO Box 332
Gaithersburg, MD 20877
(301) 948-1122

Whether or not you choose to become a member of Project Renaissance, we are most eager to hear your reactions to *The Einstein Factor.* Please feel free to send us your comments, personal stories, and insights. We want to know how these techniques have helped you and how you think they can be improved. We want to hear your own theories and personal adventures in the exciting world of brain enhancement.

With your permission, we may publish some of your letters, in full or in part, as articles or case histories in our newsletter. To facilitate handling, we suggest you mark your envelope and the top of your letter according to the following color code:

Code red: Personal letter. Not for publication.

Code orange: Can be published, in full or in part, but without attribution.

Code yellow: Can be published, in full or in part, only under the pseudonym or initials indicated.

Code green: Can be published, in full or in part, and attributed by name.

We hope you have enjoyed *The Einstein Factor.* Please stay in touch.

NOTES

Preface

1. John Patrick Zmirak, "Brain-Wave Biofeedback—a Tool for Peak Performance," *Baton Rouge Business Report*, 25 July 1995, page 74.
2. Richard J. Herrnstein and Charles Murray, *The Bell Curve: Intelligence and Class Structure in American Life*, The Free Press/Simon & Schuster, New York, 1994, page 403.

Chapter 1

1. Mihaly Csikszentmihalyi, *Flow: The Psychology of Optimal Experience*, HarperPerennial, New York, 1990, page 29.
2. "Intact Brain Filters Awareness," *Brain/Mind Bulletin*, Interface Press, Los Angeles, Calif., 4 July 1977, vol. 2, no. 16, page 2.
3. John McCrone, "Shots Faster Than the Speed of Thought," *The Independent*, 27 June 1993, page 71.
4. Michael Talbot, *The Holographic Universe*, HarperPerennial, New York, 1992, page 21.
5. Thomas G. West, *In the Mind's Eye: Visual Thinkers, Gifted People with Learning Difficulties, Computer Images, and the Ironies of Creativity*, Prometheus Books, Buffalo, N.Y., 1991, page 135.

6. Ibid., page 140.
7. Ibid., page 138.
8. Ibid., pages 118–122.
9. Ibid.
10. Cynthia Mayer, "Picking Einstein's Brain After Nearly 40 Years," *Akron Beacon Journal*, 24 October 1993, page A2.
11. Bruce Bower, "Getting Into Einstein's Brain," *Science News,* 25 May 1985, vol. 127, page 330.
12. Marian C. Diamond, Arnold B. Scheibel, Greer M. Murphy, Jr., and Thomas Harvey, "On the Brain of a Scientist: Albert Einstein," *Experimental Neurology*, April 1985, vol. 88, no. 1, pages 198–204.
13. M. C. Diamond, R. E. Johnson, A. M. Protti, C. Ott, and L. Kajisa, "Plasticity in the 904-Day-Old Rat," *Experimental Neurology*, 12 February 1985, vol. 87, no. 2, pages 309–317.
14. Marian Diamond, *Enriching Heredity*, Free Press, New York, 1988.
15. Santiago Ramon y Cajal, *Histologie du Systéme Nerveux de L'homme et des Vertébrés*, vols. I and II, Malorie, Paris: 1909–1911. (cited in José M. R. Delgado, M.D., *Physical Control of the Mind: Toward a Psychocivilized Society*, Harper & Row, New York, 1969, page 53).
16. Bower, "Getting Into Einstein's Brain," page 330.
17. Abraham Pais, *Subtle Is the Lord, The Science and Life of Albert Einstein*, Oxford University Press, New York, 1982, page 131.
18. Robert B. Dilts, *Strategies of Genius,* vol. II, Meta Publications, Capitola, Calif., 1994, pages 85–86.
19. Pais, *Subtle Is the Lord*, page 131.
20. Robert B. Dilts, *Strategies of Genius*, vol. II, pages 46–48.
21. Ibid., pages 48–49.
22. Ibid., page 48.

Chapter 2

1. Arthur Asa Berger, "Daydreaming," *Whole Earth Review*, Summer 1992, no. 75, page 37(1).
2. *Grolier's Academic American Encyclopedia*, Online Edition, Grolier Electronic Publishing, 1994, "REM sleep."
3. Thomas G. West, *In the Mind's Eye: Visual Thinkers, Gifted People with Learning Difficulties, Computer Images, and the Ironies of Creativity*, Prometheus Books, Buffalo, N.Y., 1991, page 143.
4. A. R. Luria, *The Mind of a Mnemonist,* Harvard University Press, 1968, page 25.

5. Ibid., page 24.
6. Ibid., page 82.
7. West, *In the Mind's Eye,* page 143.
8. Luria, *Mind of a Mnemonist*, page 28.
9. Ibid., pages 52–53.
10. Ibid., page 64.
11. Edward Dolnick, "What Dreams Are (Really) Made Of," *The Atlantic,* July 1995, vol. 266, no. 1, page 41(16).
12. Charles P. Reinert, "A Preliminary Comparison Between Two Methods of Intellectual Skill Development" (unpublished). Presented to the annual Conference of the Society for Accelerative Learning and Teaching (SALT), San Diego, Calif., 27 April 1989.
13. Charles P. Reinert, "A Preliminary Study of the Effect of Verbally Described Imagery in the Development of Intellectual Skills at the University Level" (unpublished). Presented to the annual Conference of the Society for Accelerative Learning and Teaching (SALT), Chicago, Ill., 27–30 April 1990.
14. Bruce Bower, "Million Cell Memories," *Science News*, 15 November 1986, vol. 130, page 313(3).
15. Francis Crick, *The Astonishing Hypothesis*, Charles Scribner's Sons, New York, 1994, pages 182–183.
16. Dolnick, "What Dreams Are (Really) Made Of," page 41(16).
17. Lawrence Miller, "REM Sleep: Pilot Light of the Mind?" *Psychology Today*, September 1987, vol. 21, page 8(2).
18. Reinert, "Two Methods of Intellectual Skill Development."

Chapter 3

1. Robert B. Dilts, *Strategies of Genius*, vol. I, Meta Publications, Capitola, Calif., 1994, pages 166–167.
2. Brad Lemley, "The Sixth Sense," *The Washington Post*, 7 July 1985, page 4.
3. Ibid.
4. Jon Van, "Brain Is Able to 'Sift' Data, Study Confirms," *Chicago Tribune*, 6 June 1985, Section 6, 7.
5. Daniel Goleman, "New View of Mind Gives Unconscious an Expanded Role," *New York Times*, 7 February 1984, page C1.
6. A. R. Luria, *The Mind of a Mnemonist*, Harvard University Press, 1968, pages 81–82.

7. John Wolkes, "A Study in Hypnosis," *Psychology Today*, January 1988, vol. 20, page 22(6).
8. Carl Jung et al., *Man and His Symbols*, Doubleday & Co., New York, 1964, page 27.
9. Robert B. Dilts, *Strategies of Genius,* vol. III, Meta Publications, Capitola, Calif. 1995, page 323.
10. Ibid.
11. William Blake, "Auguries of Innocence," *William Blake: A Selection of Poems and Letters*, ed. J. Bronowski, Penguin, New York, 1958, page 67.

Chapter 4

1. Catherine M. Cox, "The Early Mental Traits of Three Hundred Geniuses," *Genetic Studies of Genius*, vol. II, Stanford University Press, 1926.
2. Thomas G. West, *In the Mind's Eye: Visual Thinkers, Gifted People with Learning Difficulties, Computer Images, and the Ironies of Creativity*, Prometheus Books, Buffalo, N.Y., 1991, page 140.
3. Daniel Golden and Alexander Tsiaras, "Building a Better Brain," *Life*, July 1994, vol. 17, no. 7, pages 62–69.
4. Marian Diamond, *Enriching Heredity*, Free Press, New York, 1988.
5. "Brain Power (and Aging)," Joannie M. Schrof, U.S. News and World Report, 28 November 1994, vol. 117, no. 21, pages 88–87.
6. Santiago Ramon y Cajal, *Histologie du Systemé Nerveux de L'homme et des Vertébrés*, vols. I and II, Malorie, Paris, 1909–1911.
7. José M. R. Delgado, *Physical Control of the Mind: Toward a Psychocivilized Society*, Harper & Row, New York, 1969, page 49.
8. William D. Misner, "The Significance of Mobility in Early Childhood: Comparison of Two American Indian Cultures," *Human Potential*, vol. 2, no. 1, 1969.
9. West, *In the Mind's Eye*, page 29.
10. Ibid., pages 103, 112.
11. Joseph Schwartz and Michael McGuinness, *Einstein for Beginners*, Pantheon Books, New York, 1979, page 65.
12. West, *In the Mind's Eye*, pages 9–10.
13. Ibid.
14. Ibid., pages 109, 111.

15. Gary Zukav, *The Dancing Wu Li Masters: An Overview of the New Physics*, William Morrow & Co., Inc., New York, 1979, page 136.
16. For the interested reader, my economic ideas are explored in a privately published book, *Incentives as a Preferred Instrument of Corporate and Public Policy*, Project Renaissance, Gaithersburg, Md., 1995. Tel.: (800) 649-3800.
17. Alex F. Osborn, *Applied Imagination: Principles and Procedures of Creative Problem-Solving*, Charles Scribner's Sons, New York, 1953, pages 88–89.
18. Ibid., (citing Fritz Kahn, *Design of the Universe*), page 45.
19. Irvin Rock and Stephen Palmer, "The Legacy of Gestalt Psychology," *Scientific American*, December 1990, vol. 263, no. 6, page 84(7).
20. Win Wenger, "On Raising Human Intelligence: An Interdisciplinary Inquiry as to Whether Intelligence Can Be Increased," Project Renaissance, Gaithersburg, Md., 1972, 1990 edition.
21. Jill Neimark, "It's Magical. It's Malleable. It's . . . Memory," *Psychology Today*, January–February 1995, vol. 28, no. 1, page 44(8).

Chapter 5

1. Thomas West, *In the Mind's Eye: Visual Thinkers, Gifted People with Learning Difficulties, Computer Images, and the Ironies of Creativity*, Prometheus Books, Buffalo, N.Y. 1991, page 195.
2. Stephen M. Kosslyn, "Aspects of a Cognitive Neuroscience of Mental Imagery," *Science*, 17 June 1988, vol. 240, no. 4859, pages 1621–1626.
3. Michael I. Posner, Steven E. Petersen, Peter T. Fox, and Marcus E. Raichle, "Localization of Cognitive Operations in the Human Brain," *Science*, 17 June 1988, vol. 240, no. 4859, pages 1627(5).
4. Alex F. Osborn, *Applied Imagination: Principles and Procedures of Creative Problem-Solving*, Charles Scribner's Sons, New York, 1953, page 127.
5. Donald T. Phillips, *Lincoln on Leadership*, Warner Books, N.Y., 1992, page 69.
6. West, *In the Mind's Eye*, pages 197–198.
7. Ibid.

8. Ibid., page 111.
9. Julius Caesar, *The Gallic War*, Book VIII: 4, Harvard University Press, 1917.
10. Edgar D. Mitchell, Sc.D., and Robert Staretz, M.S., "Facts About Subconscious Accelerated Learning—The Natural Learning Technique," unpublished draft, 18 May 1994.
11. Irvin Rock and Stephen Palmer, "The Legacy of Gestalt Psychology," *Scientific American*, December 1990, vol. 263, no. 6, page 84(7).
12. Walter J. Freeman, "The Physiology of Perception: The Brain Transforms Sensory Messages into Conscious Perceptions Almost Instantly," *Scientific American*, February 1991, vol. 264, no. 2, page 78(8).
13. Ibid.
14. John P. Ertl, "Evoked Potentials and Intelligence," lecture delivered by the Director of Cybernetic Studies at University of Ottawa, 30 March 1967, reprinted under the title "Intelligence Testing by Brainwaves." *MENSA Bulletin*, CX, April 1968.
15. Ned Herrmann, *The Creative Brain*, Brain Books, Lake Lure, N.C., 1989.

Chapter 6

1. Carl Jung et al., *Man and His Symbols,* Doubleday & Co., New York, 1964, page 56.
2. Ibid.
3. John Horgan, "Lucid Dreaming Revisited," *Omni*, September 1994, vol. 16, no. 12, page 44(6).
4. Stephen Laberge and Howard Rheingold, *Exploring the World of Lucid Dreaming*, Ballantine Books, New York, 1990.

Chapter 7

1. John Spencer, *The Paranormal: A Modern Perspective on the Forces Within, Without and Beyond*, Crescent Books, New York, 1992, page 81.
2. Ibid., page 148.
3. Win Wenger, *How to Increase Your Intelligence*, D.O.K. Publishers, East Aurora, NY, 1987, page 70.

4. Mihaly Csikszentmihalyi, *Flow: The Psychology of Optimal Experience,* HarperPerennial, New York, 1991, page 33.
5. Robert B. Dilts, *Strategies of Genius*, vol. I, Meta Publications, Capitola, CA, 1994, page 230.
6. Ibid., page 232.
7. W. T. Jones, *A History of Western Philosophy: The Classical Mind*, Harcourt, Brace, & World, Inc., New York, 1969, page 192.
8. Ibid., page 137.
9. Jean Starobinski, "The Age of Genius," *UNESCO Courier*, July 1991, page 18(4).
10. John Horgan, "Quantum Theory," *Scientific American*, July 1992, vol. 267, no. 1, page 94(9).
11. Michael Talbot, *The Holographic Universe*, HarperPerennial, New York, 1992, pages 41–42.
12. Itzhak Bentov, *Stalking the Wild Pendulum*, Dutton, N.Y., 1977, pages 11, 14. (Cited in Peter Russell, *The Brain Book*, page 153.)
13. Peter Russell, *The Brain Book*, E. P. Dutton, Inc., New York, 1979, page 152.
14. Ellen Muraskin, "Memory Crystal," *Popular Science*, August 1992, vol. 241, no. 2, page 38(1).
15. Russell, *Brain Book*, pages 152–153.
16. Talbot, *Holographic Universe*, page 41.
17. Ibid., page 50.
18. Richard S. Broughton, Ph.D., *Parapsychology: The Controversial Science,* Ballantine Books, New York, 1991, pages 74–75.
19. Talbot, *Holographic Universe*, page 3.
20. Ibid., pages 52–53.
21. Michael Talbot, *Beyond the Quantum*, Bantam Books, New York, 1988, pages 27–39.
22. Talbot, *Holographic Universe*, page 5.

Chapter 8

1. *Superlearning* went on to sell over 1.2 million copies. Its main sequel, *Superlearning 2000*, was published by Delacorte Press in 1994. An updated version of Ostrander and Schroeder's *Psychic Discoveries Behind the Iron Curtain*, first published in 1970, is scheduled for release in 1996. For information, call Superlearning, Inc.: (212) 279–8450, Fax: (212) 695-9288.

2. Sheila Ostrander and Lynn Schroeder with Nancy Ostrander, *Superlearning 2000*, Delacorte Press, New York, 1994, pages 150–155.
3. Ward Rutherford, *The Druids: Magicians of the West,* Sterling Publishing Co., Inc., New York, 1990, page 126.
4. Charles Squire, *Celtic Myth and Legend*, Newcastle Publishing Co., Inc., Hollywood, Calif., 1975 (First edition, 1905), page 124.
5. Magaly Olivero, "Get Crazy! How to Have a Breakthrough Idea," *Working Woman*, September 1990, vol. 15, no. 9, page 144(5).
6. Michael Talbot, *The Holographic Universe*, HarperPerennial, New York, 1992, pages 76–77, 98–99.
7. Raymond A. Moody, Jr., Ph.D., M.D., "Family Reunions: Visionary Encounters with the Departed in a Modern-Day Psychomanteum," *Journal of Near-Death Studies*, 11(2), Winter 1992.
8. Jayne Gackenback and Jane Bosveld, "Take Control of Your Dreams," *Psychology Today*, October 1989, vol. 23, no. 10, page 27(6).
9. Ibid.
10. Robert B. Dilts, *Strategies of Genius,* vol. I, Meta Publications, Capitola, Calif., 1994, pages 174–175.
11. Napoleon Hill, *Think and Grow Rich*, Fawcett Crest, New York, 1960 (rev. ed.), pages 168–169, 217–219.
12. Michael Gazzaniga, *The Social Brain: Discovering the Networks of the Mind*, Basic Books, New York, 1985.

Chapter 9

1. Jung, Carl et al., *Man and His Symbols*, Doubleday & Co., New York, 1964, page 37.
2. Robert B. Dilts, *Strategies of Genius,* vol. II, Meta Publications, Capitola, Calif., 1994, page 185.
3. Thomas G. West, *In the Mind's Eye*, Prometheus Books, Buffalo, N.Y., 1991, page 124.
4. Paul Pilzer, *Unlimited Wealth: The Theory and Practice of Economic Alchemy*, Crown Publishers, New York, 1990, page 104.
5. Stephen A. Booth, "Fast Chips," *Popular Mechanics*, August 1994, vol. 71, no. 8, page 36(4).
6. Paul Zane Pilzer, interview, 24 August 1994.
7. Pilzer, *Unlimited Wealth*, page 105.

8. Constance Holden, "Academy Helps Army Be All That It Can Be," *Science*, 11 December 1987, vol. 238, no. 4833, page 1501(2).
9. Wilson Bryan Key, *Subliminal Seduction*, Signet/Penguin Books, New York, 1974, pages 22–23.
10. Ibid.
11. Paul R. Scheele, MA, *The PhotoReading Whole Mind System*, Learning Strategies Corporation, Wayzata, Minn., 1993.
12. Avi Karni, David Tanne, Barton S. Rubenstein, Jean J. M Askenasy, and Dov Sagi, "Dependence on REM Sleep of Overnight Improvement of a Perceptual Skill," *Science*, 29 July 1994, vol. 265, no. 5172, page 679(4).

Chapter 10

1. Robert Jay Lifton, *Thought Reform and the Psychology of Totalism: A Study of 'Brainwashing' in China*," University of North Carolina Press, Chapel Hill, N.C., 1989, page 3.
2. Irvin Rock and Stephen Palmer, "The Legacy of Gestalt Psychology," *Scientific American*, December 1990, vol. 263, page 84(7).
3. Robert B. Dilts, *Strategies of Genius*, vol. II, Meta Publications, Capitola, Calif., 1994, page 185.

Chapter 11

1. "Renegades: Harassed and Scorned, They Did It Their Way—and Won," *Success*, January/February 1990, page 30.
2. Donna Partow, "Great Ideas," *Home Office Computing*, June 1994, vol. 12, no. 6, page 85 (5).
3. Robert M. Palter, ed., *Toward Modern Science, vol. I: Studies in Ancient and Medieval Science*, Farrar, Straus & Cudahy, New York, 1961.
4. Marian Diamond, *Enriching Heredity*, Free Press, New York, 1988.
5. Bruce Bower, "The Great Brain Drain: A Controversial Theory Takes Ancestral Brain Growth in Vein," *Science News*, 13 October 1990, vol. 138, no. 15, page 232(3).
6. Elaine Morgan, "The Water Baby," *New Statesman and Society*, 11 December 1992, vol. 5, no. 232, page 29

Chapter 12

1. Black Elk, *Black Elk Speaks*, University of Nebraska Press, Lincoln, Nebr., 1979.

Chapter 13

1. Britt Anderson, M.D., Assistant Professor of Neurology, University of Alabama, Birmingham, interview, 22 August 1995.
2. "Musical Brains Pitch to the Left," *Science News*, 11 February 1995, vol. 147, no. 6, page 88 (1).
3. Susan Okie, "Neurology: The Asymmetry of Perfect Pitch," *Washington Post*, 6 February 1995, page A2.
4. Gottfried Schlaug, L. Jankcke, Y. Huang, and H. Steinmetz, "In Vivo Evidence of Structural Brain Asymmetry in Musicians," *Science*, 3 February 1995, vol. 267, no. 5198, page 699.
5. Peter Russell, *The Brain Book,* E. P. Dutton, Inc., New York, 1979, pages 22–23.
6. Judith Hooper, "The Boundary Factor," *The Omni WholeMind Newsletter*, November 1988, pages 1, 3.
7. Erica E. Goode, "Psychic Borderlines: A New Personality Theory Focuses on Mental and Emotional Permeability," *U.S. News and World Report*, 20 January 1992, vol. 112, no. 2, page 57(3).
8. Hooper, "Boundary Factor," pages 1, 3.
9. Goode, "Psychic Borderlines," page 57(3).
10. Maria Montessori, *Spontaneous Activity in Education*, Cambridge, Mass., Robert Bentley, Inc., 1964.
11. O. K. Moore, "On Responsive Environments," Moore's private monograph prepared for the 1964 Abington Conference, "New Directions in Individualizing Instruction." Also see Moore with Anderson, "Some Principles for the Design of Clarifying Educational Environments" in David A. Goslin, ed., *Handbook of Socialization Theory & Research*, NY: Rand McNally & Co., New York, 1969.
12. Cushman, Kathleen, "The Montessori Story," *Parents' Magazine*, January 1993, vol. 68, no. 1, page 80(5).
13. Bruce Bower, "Brain Images Reveal Cerebral Side of Music," *Science News*, 23 April 1994, vol. 145, no. 17, page 260(1).
14. Ronald W. Clark, *Einstein: The Life and Times*, Avon Books, New York, 1994, pages 140–141.

15. Thomas G. West, *In the Mind's Eye*, Prometheus Books, Buffalo, N.Y., 1991, page 120. (Quoted from: M. Winteler-Einstein, "Sketch," in *Papers*, vol. 1, 1987, page xix.)
16. Clark, *Einstein,* page 29.
17. West, *In the Mind's Eye,* page 120.
18. Allan Miller and Dorita Coen, "The Case for Music in the Schools," *Phi Delta Kappan*, February 1994, vol. 75, no. 6, page 459(3).
19. Mike Snider, "Mozart's Music May Sharpen the Mind," *USA Today*, 14 October 1993, page 1D.
20. Frances H. Rauscher, Gordon L. Shaw, and Katherine N. Ky, "Music and Spatial Task Performance," *Nature*, 14 October 1993, vol. 365, page 611.
21. Sheila Ostrander and Lynn Schroeder with Nancy Ostrander, *Superlearning 2000*, Delacorte Press, New York, 1994, page 111.
22. Miller and Coen, "Music in the Schools," page 459(3).
23. Jack Gordon, "Mainstreaming Accelerated Learning," *Training*, May, 1989, vol. 26, no. 5, page 81.
24. Ostrander, Schroeder, and Ostrander, *Superlearning 2000*, pages 69–73.
25. Ibid., page 76.
26. Ibid., pages 86–96.
27. Frederick Le Boyer, *Birth Without Violence*, Knopf, New York, 1975.
28. Bonnie Pruden, *How to Keep Your Child Fit from Birth to Six*, Ballantine Books, New York, 1986.
29. Win Wenger and Susan Wenger, "Training Music Sight-Reading and Perfect Pitch in Young Children, As a Way of Enhancing Their Intelligence," *Journal of the Society for Accelerative Learning and Teaching*, 1990, 15 (1 and 2).

Chapter 14

1. Col. John B. Alexander, Maj. Richard Groller, and Janet Morris, *The Warrior's Edge*, William Morrow and Company, Inc., New York, 1990, page 92.
2. "Weapons, Deadly Weapons," *The Economist*, 18 February 1989, vol. 310, page 18
3. Karl W. Luckert, *Egyptian Light and Hebrew Fire: Theological and Philosophical Roots of Christendom in Evolutionary Perspective*, State University of New York Press, Albany, 1991, page 101.

4. Lev Vygotsky, *Thought and Language*, M.I.T. Press, Cambridge, Mass., 1962.
5. Ronald W. Clark, *Einstein: The Life and Times*, Avon Books, New York, 1994, page 30.
6. Mihaly Csikszentmihalyi, *Flow: The Psychology of Optimal Experience,* HarperPerennial, New York, 1990, page 1.
7. Joannie M. Schrof, "Brain Power," U.S. News and World Report, 28 November 1994, vol. 117, no. 21, page 88(7).
8. Edwin M. Reingold, "Mathematics Made Easy: A Japanese Teaching Method Adds Up in the U.S.," *Time*, 4 June 1990, vol. 135, no. 23, page 83(2).
9. Robert B. Dilts, *Strategies of Genius*, vol. 1, Meta Publications, Capitola, Calif. 1994, page 176.
10. Ibid., page 169.
11. Jeff Olson, "The Slight Edge," *Success*, January/February 1995, page 20.
12. Clark, *Einstein*, page 85.
13. Csikszentmihalyi, *Flow,* page 67.
14. Robert B. Dilts, *Strategies of Genius*, vol. II, Meta Publications, Capitola, Calif. 1994, page 198.

Chapter 15

1. Kurt Mendelssohn, *The Riddle of the Pyramids*, Praeger Publishers, New York, 1974, page 40.
2. Robert B. Dilts, *Strategies of Genius,* vol. II, Meta Publications, Capitola, Calif. 1994, page 18.
3. Howard Bloom, *The Lucifer Principle: A Scientific Expedition into the Forces of History*, Atlantic Monthly Press, New York, 1995.
4. Ibid.
5. Keith Henson, "Memetics: The Science of Information Viruses," *Whole Earth Review*, Winter 1987, no. 57, page 50(6)
6. Bloom, *Lucifer Principle*, page 101.
7. Mendelssohn, *Riddle of the Pyramids*, page 50.
8. Donald T. Phillips, *Lincoln on Leadership*, Warner Books, New York, 1992, page 108.
9. Napoleon Hill, *Think and Grow Rich*, Fawcett Crest, New York, 1960, page 184.
10. Walter J. Freeman, "The Physiology of Perception: The Brain Transforms Sensory Messages into Conscious Perceptions Almost

Instantly," *Scientific American*, February 1991, vol. 264, no. 2, page 78(8).
11. Luiz Machado, *The Brain of the Brain*, Cidade do Cérebro, Brazil, 1990, page 141.
12. Freeman, "Physiology of Perception," page 78(8).
13. Kenneth M. Heilman, and Paul Satz, *Neuropsychology of Human Emotion*, Guilford Press, New York, 1983, page 93.
14. Hainer, R. "Rationalism, pragmatism and existentialism," in E. Glatt and M. Shelley, *The Research Society,* Gordon and Breech Science Publications, New York, 1968. (Cited in Luiz Machado, *The Brain of the Brain*, Cidade do Cérebro, Brazil, 1990, pages 56–57).
15. Ibid.
16. Dilts, *Strategies of Genius*, vol. II, page 10.
17. Ibid., page 7.
18. Christy G. Turner, Regents Professor, Arizona State University, Tempe, Ariz., interview, 22 August 1995.
19. Peter Russell, *The Brain Book,* E. P. Dutton, Inc. New York, 1979, page 105.
20. Henson, "Memetics," page 50(6).

INDEX

A

B

C

D

E

H

I

P

Q

R

S

T

U

V

W